BUILD-A-BOT

Use 3DEXPERIENCE SOLIDWORKS for Makers to bring your next competition robot to life!

When he's not busting IT vulnerabilities, Chicago-based information security professional **Brian Adamson** is designing, building, and competing in *BattleBots* events.

His interest began while growing up in Alaska, watching the early 2000s episodes of the show with his brother. "It always seemed like so much fun, and the community of builders was, as it is now, generally very nice and welcoming," he says.

After moving to the Midwest and discovering the Central Illinois Robotics Club (CIRC) in Peoria, Adamson decided to get involved in their robot league. "I taught myself CAD, made a ridiculous number of mistakes, was taught by people smarter and more experienced than me, and built my first 3lb robot," he says. "I even took second place my first year!"

He followed up with a first-place finish in his second year, and then embarked on more, and larger, robot builds. His most recent creation, Claw Viper, brought him to the 2021 season of that show that originally inspired him. "I designed that machine, from scratch, using SOLIDWORKS, and it was actually my first large project in SOLIDWORKS," Adamson says. "It's a 250-pound monster with over 50 horsepower — more than 40kW — of drive power alone, it makes over 1,000 foot-pounds of torque at the main lift axle, and it has the additional craziness of a couple hundred pounds of additional magnetic downforce on the chassis to make it possible to get all that power to the ground."

Adamson is now underway on his next entry, opting to use SOLIDWORKS' new low-cost CAD tool for it. "There are a lot of different options out there for doing design work, and I've been having a great time using **3D**EXPERIENCE SOLIDWORKS for Makers on my latest combat robotics project," he says. "SOLIDWORKS is an incredible tool, and putting it in the hands of hobbyists and makers is tremendous — it offers real, industry-leading software to anyone with a passion for creating, and creates an incredible pipeline for people to make their dreams a reality."

3DEXPERIENCE SOLIDWORKS for Makers is now available for anyone doing personal, non-commercial projects. For just US $99/year or US $9.99/month you'll get the same intuitive cloud-connected CAD modeling tools that the professionals use, along with:

- Fully online design solutions you can access from any web browser — no download required
- An online community that lets you connect with worldwide makers from fab labs, makerspaces, and influencers, all ready to share their designs, ideas, and expertise
- Access to an expanded professional ecosystem to rapid prototype your parts, or receive engineering services via the **3D**EXPERIENCE Marketplace
- Support to help you get the most out of **3D**EXPERIENCE SOLIDWORKS for Makers

CONTENTS Make: 81

22

Cover photo: Jon C R Bennett / JCRBPhoto

34

56

78

84

96

108

120

Jon C R Bennett / JCRBPhoto, Tomás Vega, Eyal Perry, Adam Haar, Abhi Jain, Ivana Hackova, Kevin Webb, Brian Mernoff, Bob Knetzger, Ship Finder / shipfinder.co

Make:

> "[Like NASCAR] it's the crashes, fires, and explosions that make the highlight reel — which is what robot sports are all about!" —*David Calkins*

PRESIDENT
Dale Dougherty
dale@make.co

VP, PARTNERSHIPS
Todd Sotkiewicz
todd@make.co

EDITORIAL

EXECUTIVE EDITOR
Mike Senese

SENIOR EDITORS
Keith Hammond
keith@make.co
Caleb Kraft
caleb@make.co

PRODUCTION MANAGER
Craig Couden

CONTRIBUTING EDITOR
William Gurstelle

CONTRIBUTING WRITERS
Massimo Banzi, Jon C R Bennett, Benjamin Cabé, David Calkins, Emmanuel Carrillo, Kathy Ceceri, David Covarrubias, Tim Deagan, Lucy Du, Peter Garnache, Greg Gilman, Adam Haar, Ivana Huckova, Abhi Jain, Bob Knetzger, Caleb Kodama, Sophie Martinez, Austin McChord, Brian Mernoff, Forrest M. Mims III, onemindisbuddha, Eyal Perry, Marshall Piros, Charles Platt, Oscar Roselló, Bunny Sauriol, Seth Schaffer, Michael Shiloh, Tomás Vega, Kurtis Wanner, Kevin Webb, Lee Wilkins, Brandon Bennett Young

CONTRIBUTING ARTIST
Jon C R Bennett

MAKE.CO

ENGINEERING MANAGER
Alicia Williams

WEB APPLICATION DEVELOPER
Rio Roth-Barreiro

DESIGN

CREATIVE DIRECTOR
Juliann Brown

BOOKS

BOOKS EDITOR
Michelle Lowman
books@make.co

GLOBAL MAKER FAIRE

MANAGING DIRECTOR, GLOBAL MAKER FAIRE
Katie D. Kunde

GLOBAL LICENSING
Jennifer Blakeslee

MARKETING

DIRECTOR OF MARKETING
Gillian Mutti

COMMUNITY MANAGER
Dan Schneiderman

OPERATIONS

ADMINISTRATIVE MANAGER
Cathy Shanahan

ACCOUNTING MANAGER
Kelly Marshall

OPERATIONS MANAGER & MAKER SHED
Rob Bullington

PUBLISHED BY

MAKE COMMUNITY, LLC
Dale Dougherty

Copyright © 2022 Make Community, LLC. All rights reserved. Reproduction without permission is prohibited. Printed in the USA by Schumann Printers, Inc.

Comments may be sent to:
editor@makezine.com

Visit us online:
make.co

Follow us:
🐦 @make @makerfaire @makershed
📘 makemagazine
📷 makemagazine
▶ makemagazine
🎮 twitch.tv/make
📌 makemagazine

Manage your account online, including change of address: makezine.com/account
For telephone service call 847-559-7395 between the hours of 8am and 4:30pm CST.
Fax: 847-564-9453.
Email: make@omeda.com

Make: Community

Support for the publication of *Make:* magazine is made possible in part by the members of Make: Community. Join us at make.co.

CONTRIBUTORS

If you could pilot any robot/mech, real or imagined, what would it be and why?

Brandon Bennett Young *Bowie, MD*
(Build Your First Combat Robot)
I would pilot the Battlebot "Hydra." I would love sending things flying with it!

Caleb Kodama
Pasadena, CA
(Sick Sniffer)
I'd pilot a Titan from the game *Titanfall*. They are the best versatile mechs for medical, farming and warfare applications.

Lucy Du
Cambridge, MA
(The Roboteers)
Iron Man suit. As a designer and builder of high-tech prostheses, it would be awesome to pilot the ultimate high-tech prosthesis.

Issue No. 81, Summer 2022. *Make:* (ISSN 1556-2336) is published quarterly by Make Community, LLC, in the months of February, May, Aug, and Nov. Make: Community is located at 150 Todd Road, Suite 100, Santa Rosa, CA 95407. SUBSCRIPTIONS: Send all subscription requests to *Make:*, P.O. Box 566, Lincolnshire, IL 60069 or subscribe online at makezine.com/subscribe or via phone at (866) 289-8847 (U.S. and Canada); all other countries call (818) 487-2037. Subscriptions are available for $34.99 for 1 year (4 issues) in the United States; in Canada: $43.99 USD; all other countries: $49.99 USD. Periodicals Postage Paid at San Francisco, CA, and at additional mailing offices. POSTMASTER: Send address changes to *Make:*, P.O. Box 566, Lincolnshire, IL 60069. Canada Post Publications Mail Agreement Number 41129568.

HAVE A BLAST
WITH OUR
COMPRESSED AIR ROCKET LAUNCHER V2.2

- Blast your paper rockets hundreds of feet high!
- Newest version with quick 15-minute assembly
- Sturdy design for many years of use
- Includes rocket kit, spare foam nose, 5 paper templates & sticker
- Ships FREE

Go even further with extra rocket sets and Make: book *Planes, Gliders, and Paper Rockets*

Make:
Planes, Gliders, and Paper Rockets

Simple Flying Things Anyone Can Make—Kites and Copters, Too!
Rick Schertle & James Floyd Kelly

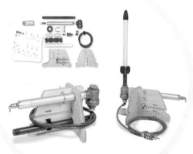

Maker Shed
makershed.com

Make: KITS FOR YOUNG MAKERS

Make:Kits

Papercraft Circuits
Make Your Projects Shine!
makezine.com

Make:Kits

4 pk

Brush-E Robo Racers
makezine.com

Make:Kits

Origami Robot Paper Circuits
Light up Any Project!
makezine.com

Use LEDs, motors, conductive tape, and more to create easy, fun projects with these kits

- **Papercraft Circuits**
- **Brush-E Robo Racer**
- **Origami Robot Paper Circuits**

Adobe Stock-Ms.Moloko

Performing Live with Robots

by Dale Dougherty, President of Make: Community

The moment Mario the Maker Magician walked on stage at the Sebastiani Theater in Sonoma, California, every single child in the audience followed his every move and every word. For over an hour, he performed magic, but his real magic was how the children were beside themselves in wonder and laughter.

I sat in the second row with my two young grandchildren, surrounded by many more kids bouncing up and down in their seats. I wished every kid could experience Mario's performance, especially now as we struggle to come out of Covid, and we realize what we've been missing. Kids need to experience being a part of live events.

In talking to Mario, I realized he needed this, too, to use his special talent. For the last two years, he hasn't been able to do what he does best. He adapted, just like all of us during Covid. He performed for an audience in front of a camera but he was really by himself, just as each person watching him was by themselves. It wasn't the same as creating the energy and excitement with a live audience.

Earlier this year, Mario and Katie Marchese decided it was time to go out on tour again. Their two children, Gigi and Bear, would go with them and help as crew. "The main goal of this tour is just to reconnect with people in all these venues," said Katie, who manages everything. Usually their caravan is a 1971 VW bus but it broke down for the last time. So they drove an SUV from New York to Georgia, and then to Tennessee, Texas, and Colorado before making it to the West Coast. "Sometimes we're asking ourselves 'What are we doing?' but for these crazy experiences, we're all together," said Katie.

Mario makes all the props he uses in the show, and he tells kids that they can make anything using stuff all around them. His props use Arduinos and one of them is a robot, which has a mind of its own. He tells the robot to do something and the robot does the opposite. Everybody laughs. You just can't expect robots to do what you want...

After shows in Sonoma and Oakland, they drove to Portland, Oregon, for a performance. The next night outside their hotel, someone broke into their SUV, stealing a bag of their props, including that adorably defiant robot, as well as some camera equipment. They also took a "Do What You Love" banner that Mario and family fly at the end of their show. The unexpected happens.

"When you get robbed, guess what? The community jumps in," said Mario with gratitude. "They reach out. They 3D-printed the parts that I lost and overnighted them to me." As they headed to Utah, Mario was building circuits in the car. "By the next show I had my robot back," said Mario. "I also had a great new story to tell about the maker community."

In every show, Katie says, Mario is telling kids that whatever toy you have in the world, "It can break. You can lose it. Someone can take it from you. But if you learn how to make something, no one can take that from you, ever." ◉

Dale Dougherty

MADE
ON EARTH

Backyard builds from around the globe

Found a project that would be perfect for Made on Earth?
Let us know: *editor@makezine.com*

THE HEXANAUT

YOUTUBE.COM/MATTDENTON

Whether they realize it or not, fans of *Star Wars*, *Harry Potter*, and the Disney+ series *Andor* may already be familiar with UK creator **Matt Denton's** movie magic. But in 2012, Denton's futuristic fabrications transcended the silver screen and landed in the real world with the debut of *Mantis* — a six-legged, two-ton, human-piloted, steel behemoth.

Originally commissioned for commercial use, a company noticed the smaller hexapods Denton was creating in the 2000s and requested a 400-ton version to explore underwater seabeds. Because of the scale, Denton was funded for Mantis to test this tech at 2 tons. Operated over Wi-Fi or using the two onboard three-axis joysticks and 28 buttons inside the cockpit, Mantis' top speed is 1km/h and, despite its weight, exerts the same pressure as a human foot under each footpad.

Mantis took over three years to build with Denton tackling the work alone for the first eight months. Denton recalls it "was a challenge for me on a daily basis mentally. I had never done hydraulics that big before and certainly never a hydraulic power pack running on a diesel engine or controlling 18 actuators simultaneously." Each leg joint was particularly difficult with Denton using twin bearings driven by a linear ram that pivots each joint. Afterwards, the team found rotary actuators that could have made Mantis' legs a single, strong unit. "I was learning on the job...even in those three years, I found better ways to do the mechanics." It's "a sort of Frankenstein thing," joked Denton. "It's a 486 processor clocked to a gigahertz, but running a Linux system." The result is a breathtaking, engineering feat earning Denton the Guinness World Record for "Largest Rideable Hexapod Robot" in 2017.

With the processing power of Arduinos and the possibilities of 3D printing, Denton has lately been exploring giant 3D-printed versions of classic Lego vehicles. To Denton, these "have changed everything" and he hasn't counted out using his new skills to tackle hexapods again in the future. Follow more of his latest builds at instagram.com/mantisrobot. —*Sophie Martinez*

Michael Hughes

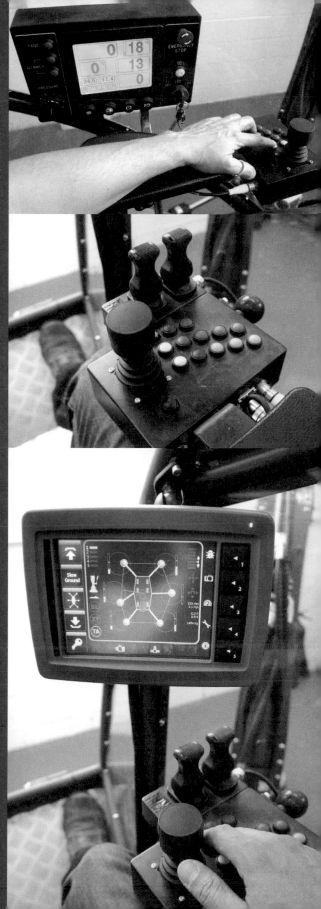

CHROME AND LACE

RAERIPPLE.COM

Meet *Phillis Gene*, a vintage Chevrolet Deluxe that just got a one-of-a-kind redesign by Texas-based metal artist **Rae Ripple**.

"I wanted to keep that real classy lady feel, so that's where the lace came in," Ripple says of her work, commissioned by a rat rod enthusiast and named after his mother. "I guess if you were going to put a piece of lace up against a woman's body, that's exactly where that inspiration came from."

Phillis is Ripple's fifth custom car, and one of dozens of art projects she's created since picking up a welding torch eight years ago. The hobby turned out to be a saving grace, sparking a creative process that would transform a life once defined by struggle, including homelessness as a teenager, into one of joy, gratitude, and purpose. Now 36, Ripple has a flourishing business and family, as seen on her vibrant Instagram feed followed by 157,000 people eager to see her next make — or best advice. "Today is a beautiful day to chase your dreams. So just go for it," reads a recent caption.

Her triumphant story and optimistic outlook compliment her art, leading to speaking engagements intertwining all three into community service to inspire the next generation of welders. Ripple just spoke at the Pennsylvania College of Technology when we connected to discuss the challenges of crafting Phillis. Using a Hypertherm 65 SYNC air plasma cutter, she free-handed all of the cuts: "I don't really draw anything on or trace anything out, I just start at one end and work my way around. So, if you overcut, there is no fixing that."

Mistakes become part of the art. Or as Ripple reframes, "There's no mistakes in art." It's a perfect metaphor for life: Sometimes we overcut, but those accidents are crucial moments in the creation of our best selves. —*Greg Gilman*

Rae Ripple

MODEL MAINFRAME

MINIATUA.COM

With how important desktop and laptop computers are to the world today, it's hard to imagine that, not too long ago, they used to be the size of refrigerators and entire rooms. However, **Nicolas Temese** has found a way to keep the memory of these gigantic machines alive — by shrinking them down to a fraction of today's size via polystyrene scale miniatures.

Temese, who lives in Montreal, Quebec, has always been fascinated by the designs and aesthetics of the first generation of computers, especially those made by tech pioneer IBM. Although he has prior experience with miniature modeling, Temese said his tiny versions of the IBM 1401 (1:15 scale) and IBM 704 (1:16 scale) took hundreds of hours over several months. The miniatures are approximately 4 inches tall, but he strove to maintain every minute detail of IBM's groundbreaking machines, 3D printing them in resin or etching them in copper for greater accuracy. The 1401 model also contains some extra non-polystyrene elements to make it feel more alive; a custom ATmega board spins the tape drives with tiny motors and controls small LEDs inside the control panels, animated pseudo-randomly to increase that sense of realism. "It's not just about making a replica," he admitted. "It's about capturing the essence of the computer and the people who built, designed, and engineered these machines."

Temese's commitment to the spirit of his miniatures caught the attention of IBM themselves, who commissioned him to make similar replicas of the modern z15 microprocessor and the Power 10 E1080 server. Additionally, IBM also greenlit a limited run of 40 miniature IBM 5150 Personal Computers to celebrate the original model's 40th anniversary. Beyond this series, Temese is still determining what on his long list of miniature projects he'd like to make next. Whatever the case, it's clear that this maker's skills are giving credence to the phrase "small but mighty." You can check out all of his past and future projects under the name "miniatua" on Instagram and YouTube, or his website miniatua.com. —*Marshall Piros*

Nicolas Temese

Get a Grip!

OVERCOME THOSE SHAKY HANDS WITH THESE USEFUL ADAPTIVE DEVICES

Written by Charles Platt

CHARLES PLATT is the author of the bestselling *Make: Electronics*, its sequel *Make: More Electronics*, the *Encyclopedia of Electronic Components Volumes 1–3*, *Make: Tools*, and *Make: Easy Electronics*. makershed.com/platt

Some of us are blessed with manual dexterity, but others, not so much. I've known people whose hands are so steady they could do brain surgery, but personally I have difficulty removing a splinter, and this puts me at a disadvantage in the maker universe.

Ever since I was a kid, I've had a mild case of "essential tremor." This is a relatively common neurological condition in which the brain over-corrects in response to feedback from the fingers, somewhat like a closed-loop servo system doing "hunting" oscillations. Unlike Parkinson's disease (which has serious consequences), essential tremor is harmless. It just annoys me when I can't apply the tip of a soldering iron precisely where it is supposed to go.

Recently I discovered that there are simple ways to deal with this problem, and the strategies and gadgets may be helpful for detail work, even if you don't have especially shaky hands.

ADDING RESISTANCE

The simplest option is to add mass to the tool that you're using, because mass has inertia, and inertia provides resistance to motion. Suppose you are lifting a dumbbell. The weight is so heavy, it will subdue any tremor in your hand. At the other extreme, if you're holding disposable wooden chopsticks that weigh a couple of grams, you may have difficulty picking up grains of rice. Even a normal steel fork is easier to use than chopsticks — and in fact people with significant tremors can buy weighted knives, forks, and spoons to improve their coordination. A set of these "adaptive utensils" is shown in Figure Ⓐ.

Other options also exist:
- If your handwriting is shaky, you can try a weighted pen holder (Figure Ⓑ).
- The all-purpose weighted handle in Figure Ⓒ could receive a toothbrush ... or a stylus for a graphics tablet.

I might be able to address my soldering-iron problem by wrapping something heavy around the handle, such as 3M weighted tape, which is sold for balancing automobile wheels. This is expensive, though, and there's a more general solution which works with almost any small hand tool. Figure Ⓓ shows a double-walled cloth pad

Adobe Stock - cloud7days, Michelle Lowman, Kinsman Enterprises, SP Ableware, Charles Platt

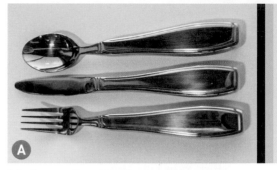

Adaptive utensils, weighted to suppress hand tremors, could be a model for precision hand tools.

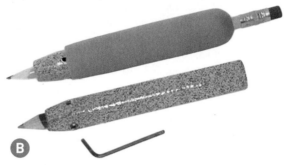

These holders are secured to pen or pencil using a grub screw that is tightened with an Allen wrench (provided).

The Maddak Universal Handle, unfortunately not large enough to accommodate the handle of a soldering iron.

The Hand Weight™ won't win any awards for design elegance, but it works.

> "I decided to buy a couple of **Hand Weights**, and they actually do make my hands steadier. They also have a beneficial psychological effect, because I don't get so frustrated as a result of struggling with my poor coordination."

filled with steel shot, which attaches to the back of your hand using a velcro wrist strap and elastic loops that fit over your fingers.

This Hand Weight (the term is trademarked) was invented by Mary Ann Heinz, an occupational therapist who started a business named Handithings 22 years ago. Her son Brian now owns the company, and I was so intrigued by the product, I contacted him in Moscow Mills, Missouri, for a quick Zoom interview.

"Initially my mother worked with special-needs children who had difficulty using their hands," he explained. "But we found that sales of larger sizes of Hand Weights were going through the roof, and I started getting feedback from hand surgeons that people were using them for Parkinson's and essential tremors. It's now our number-one product."

Handithings has provoked competition from medical equipment manufacturers, but Brian says he's happy to remain a family business whose products are made locally. "The people who sew them for us are five minutes from here," he says.

I decided to buy a couple of Hand Weights, and they actually do make my hands steadier. They also have a beneficial psychological effect, because I don't get so frustrated as a result of struggling with my poor coordination.

SOFT TIPS

My experience prompted me to revise many of my habits relating to detail work. When trying to grab a diode or a 3mm LED, I don't fumble around with my finger and thumb anymore. I use very sharp-tipped pliers such as those in Figure **E**, which make me less likely to bend component leads when I insert them into a breadboard. My favorite pliers came from Michaels, the chain of crafts stores, which sells them for making jewelry.

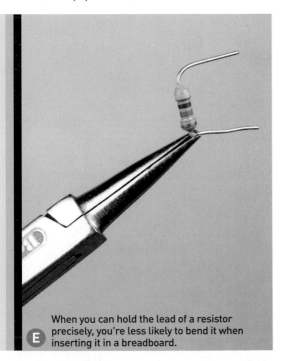

E When you can hold the lead of a resistor precisely, you're less likely to bend it when inserting it in a breadboard.

F Assorted tweezers with chisel-pointed tips, found on Amazon.

Charles Platt, Akaki Kuumeri

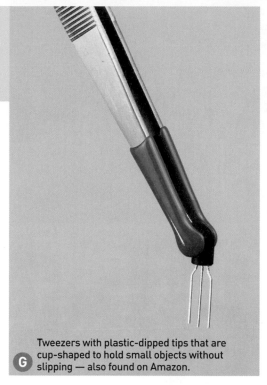

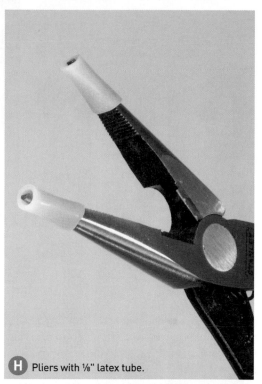

G Tweezers with plastic-dipped tips that are cup-shaped to hold small objects without slipping — also found on Amazon.

H Pliers with ⅛" latex tube.

In the past, I avoided tweezers because the tips were too thin — but when I searched online, I found tweezers with chiseled tips, shown in Figure **F**. Then I found others with cup-shaped tips coated in soft plastic, ideal for holding small, smooth objects such as transistors, as in Figure **G**.

This gave me the idea of adding compliant material around the jaws of regular long-nosed pliers. I bought ⅛" internal-diameter latex tube from McMaster-Carr, and mounted a couple of pieces on pliers as in Figure **H**. If you'd prefer not to spend money on three feet of latex when you only need an inch of it, you can try using electrical tape, although personally I found that it doesn't work as well.

GOING FURTHER
To achieve total hand stability, I can imagine a device mounted on the back of each hand using accelerometers to sense small motions and a haptic output that generates negative feedback. This would be just the thing for people who have difficulty pulling out splinters — and maybe even for brain surgeons, too. ◉

Universal Access

Makers, with their broad knowledge and range of skills, have been producing accessible devices for many years. Here are a few individuals and groups that are helping let everyone participate in all types of activities. *—Caleb Kraft*

- **AbleGamers** — ablegamers.org
- **SpecialEffect** — specialeffect.org.uk
- **Makers Making Change** — makersmakingchange.com
- **Warfighter Engaged** — warfighterengaged.org
- **Tikkun Olam Makers** — tomglobal.org/about
- **e-Nable** — enablingthefuture.org
- **The Controller Project** — thecontrollerproject.com
- **Ben Heck** — benheck.com

SICK SNIFFER

I BUILT AN AI NOSE THAT "KNOWS" FUNGAL PNEUMONIA FOR MY 8TH GRADE SCIENCE FAIR

Written by Caleb Kodama

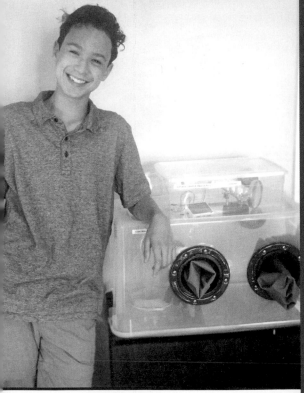

CALEB KODAMA is a 9th grader from Pasadena, California, who aspires to help the world through science and engineering. In 2020 he won 1st place in the Los Angeles County Science & Engineering Fair and Top 300 in the national Broadcom MASTERS.

I was born two months early, spending the first month of my life in the hospital. The doctors told my mother that I would probably have physical and cognitive delays. Bringing me home, my mom decided I would never be treated any differently because others told her something couldn't be done about the challenges I could face in the future. As I grew, she never once held back, not when I needed speech therapy at age 3, nor when she was told that I was eight months behind my peers in 4th grade. No matter the obstacles that were put in front of me, she persevered and kept moving forward. She always played puzzle games with me and challenged me with riddles. I didn't know it at the time, but these riddles and games like *Myst* and *The Room* helped me learn to look at the big picture while simultaneously seeing the smallest obscure details. My mother's perseverance taught me that no matter what, there is always a way.

Later in my childhood, I got pneumonia. It was terrible, and unfortunately, I was misdiagnosed because I barely had symptoms. On our first trip to the hospital, I was sent home with antibiotic ear drops. Days later we returned, and my mother pushed the point with my doctor. Once I was finally diagnosed with pneumonia, which was almost a week after I got sick, they didn't know the type. Generally, most people are diagnostically assumed to have bacterial pneumonia and given antibiotics; they gave me three different types over eight days. My fever just wouldn't quit, and on the eighth day, I was scheduled to be admitted to the hospital. Thankfully my fever finally broke just beforehand, and I spent the next month recovering. We never did determine what type of pneumonia I had.

AN IDEA IS BORN

Pneumonia had a profound effect on my life, and I began to wonder if I would have gotten sicker had it not been for my mother's unrelenting persistence. Getting sick also helped plant a seed. I just couldn't get the idea out of my head that there has to be an easier way to help people determine the type of pneumonia someone has. I had wondered about this for four years, and

now, working on my 8th grade science fair, I thought I finally might have a chance to crack it.

I JUST COULDN'T GET THE IDEA OUT OF MY HEAD THAT THERE HAS TO BE AN EASIER WAY.

To start, I began speaking to my aunt and uncle, who are both doctors. Initially, I simply wanted to identify the type of pneumonia faster, without having to rely on a blood test, x-rays, sputum samples, bronchial lavage, or the industry gold standard: gas chromatography mass spectrometry (GC/MS). Could I develop something like a breathalyzer for disease?

My aunt and uncle referred me to a pulmonologist colleague of theirs who had just so happened to be part of a study conducted on detecting invasive aspergillus, or fungal pneumonia, through patients' breath. The study focused on identifying the metabolic signatures of the disease using a GC/MS, and interestingly the chemical compounds associated were all monoterpenes/terpenes. These are derived from plant essential oils, which was perfect! This meant I could use safe, easily accessible substances such as citrus oil (limonene) and nutmeg oil (camphene) for my testing, and overcome the challenges of using potentially harmful cultures of fungal pneumonia and acquiring the aid of someone who is authorized to deal with such samples.

THE NOSE KNOWS

I remembered reading, through my research, how dogs could detect cancer by scent. The idea of scent then led me to electronic noses, or E-Noses. There have been a few studies done on other diseases like renal disease, which affects the kidneys, and diabetes. The studies focused on making a handheld device with a large enough sensory array to detect those diseases on the breath. One study used a Raspberry Pi, and it gave me hope. Maybe I could build my own E-Nose to detect fungal pneumonia, and even better, maybe I could make my E-Nose wireless.

I spent almost five long months doing

research, and I pivoted so many times during this period to keep the dream alive of quickly diagnosing fungal pneumonia — and possibly other respiratory diseases. Nothing would get in my way. By accident, I found an article in *Make:* Volume 77, by a gentleman named Benjamin Cabé. Benjamin had created an E-Nose to detect sourdough starter, whiskey(s), and coffee. The

ONCE SAMPLING WAS COMPLETE MY E-NOSE HAD A 96.5% ACCURACY BEFORE DEPLOYMENT AND 87.7% OPTIMIZED DEPLOYMENT.

E-Nose information in this article was priceless because it gave me the opportunity to use Benjamin's framework to build my own E-Nose.

I cross-referenced several studies, including Benjamin's original prototype, and determined that the ideal sensor array for my project would include Seeed Studio's multichannel gas sensor, with nitrogen dioxide (NO_2), carbon monoxide (CO), ethyl alcohol (C_2H_5OH), and volatile organic compounds sensors, and potentially a more extensive sensory array including Seeed's MQ9 (carbon monoxide, flammable gases), MQ4 (methane), MQ135 (ammonia, benzene, NOx), MQ8 (hydrogen), TGS2620 (alcohols, organic solvents), MQ136 (sulfur dioxide), TGS813 (hydrocarbons), TGS822 (organic solvents), and MQ3 (alcohols).

The E-Nose build would utilize the Seeed Wio Terminal (2.4" LCD screen, ATSAMD51 core, and Realtek RTL8720DN radio module with BLE 5.0 and Wi-Fi 2.4GHz/5GHz), utilizing its I^2C interface, MOSFET fan control, Seeed-based sensor arrays, and Seeed expansion battery pack to enable wireless connectivity.

MY PROCEDURE

I felt it was important to keep this as close to a human patient study as possible. I constructed a cleanroom out of a storage container, complete with gloves and a mechanical artificial lung that would breathe in the essential oil sample then breathe it out into another case that contained the E-Nose for sampling.

My E-Nose prototype and various sensors.

My DIY cleanroom with mechanical artificial lung.

Heather Kodama (aka Mom)

Next, and most importantly, cloning the framework of Benjamin's project I interfaced my E-Nose with Edge Impulse. Edge Impulse is a platform that takes the data you collect through supported microcontrollers and builds an artificial intelligence (AI) that can be deployed back into your microcontroller. Effectively Edge Impulse allowed the AI to "smell" through the connected sensors by identifying the four distinct monoterpenes/terpenes of fungal pneumonia, separately and mixed. Though each sample took only minutes, the combination of samples added up to hours of sample taking for each chemical compound, to aid the AI in differentiating between each chemical. Once sampling was complete my E-Nose had a 96.5% accuracy before deployment and 87.7% optimized deployment.

One of the most important components of this experiment was to make this E-Nose accessible online in real-time. Benjamin's project also included the ability to do this. Updating the necessary firmware to the Seeed Wio Terminal, I created an account on Microsoft's Azure IoT platform. Then I had a Zoom call with Benjamin, who is based in France.

Benjamin downloaded my E-Nose framework

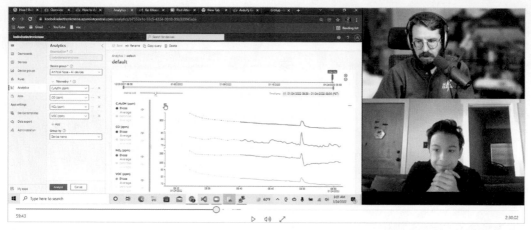

Real-time readings from my and Benjamin Cabé's E-Noses.

and connected his own E-Nose to my Azure IoT platform. In real-time both of our E-Nose frameworks started to report their readings: they were both "smelling" ambient air and indicated it corresponded to "Caleb's room."

This is incredibly important because this means doctors could have the opportunity of helping diagnose an ill patient who may not have the means to get to a hospital or a doctor's office. Even better, regarding the Covid pandemic, as new strains develop doctors can diagnostically record data faster and identify different variants more efficiently without waiting for lab test results.

MOVING FORWARD

This March I entered my project in the 2022 Los Angeles County Science Fair. I took second place! Next up, I will be competing nationally later this spring, and will be registering for international competitions too.

Granted, my E-Nose has a long way to go. This was just my first prototype. Better calibration and more-sensitive sensors will be needed — specifically, sensors that can also differentiate between the massive amount of volatile organic compounds in human breath.

Nonetheless, it is a great starting point. By joining highly sensitive sensors with AI and tiny machine learning (tinyML) to detect breath markers, in the very near future we may be able to breathe into our phones and have doctors diagnose our health issues remotely. I'm excited about what the future may bring in medical diagnostics. ◑

A NOTE FROM BENJAMIN

I built the prototype of my infamous "artificial nose" in just a couple hours (*Make:* Volume 77, makezine.com/projects/second-sense-build-an-ai-smart-nose). As a software engineer and electronics enthusiast, I was convinced that I hadn't invented anything new. Surely, "E-Nose" technology had to be available already if I could pull this together using cheap gas sensors and with zero machine-learning knowledge. I could not have been more wrong!

In retrospect, I'm glad I took the time to share this project with the world, as it has inspired many folks. For example, I had discussions with flavorists (yes, it's a thing!) who knew about smell sensors but hadn't realized what AI could bring to their field, and also with AI experts who had no idea about the landscape of available sensors. I experienced first-hand how much bias there is in the assumptions we tend to make about how others perceive a particular technology area.

When Caleb Kodama contacted me last year, I initially didn't take the time to return his emails (sorry, Caleb!). As you see from his story, he's not the kind of kid that takes silence for an answer, so he wrote to me until I finally replied. We eventually got on the phone, and as I learned more about his project, I was blown away by his creativity and what he had accomplished. His project is to me the perfect example of how the maker spirit — and open source in general — truly democratizes access to technology, and how it empowers anyone to innovate. *—Benjamin Cabé*

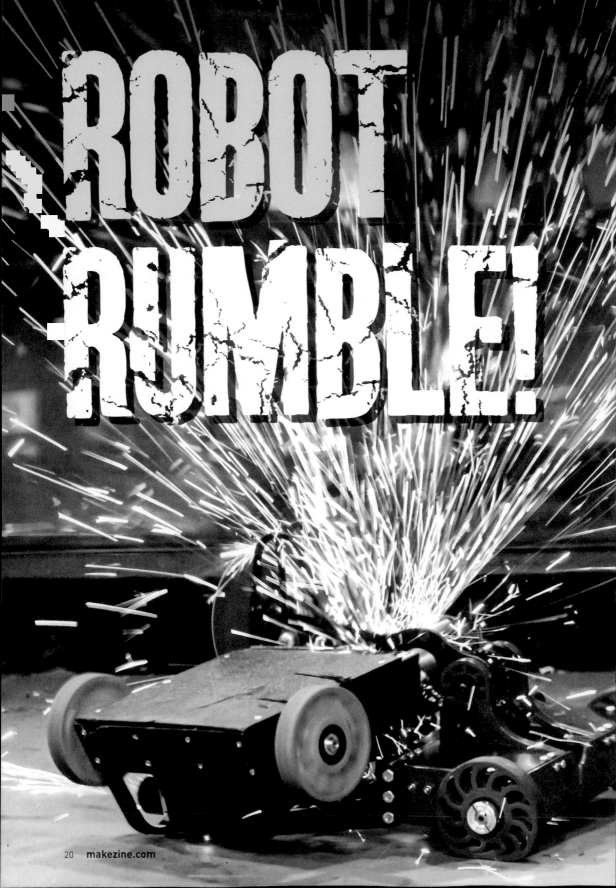

ROBOT RUMBLE!

CONTENTS

Put your bot-building and arena-navigating skills to the test by going head-to-head with your best pals in radio-controlled, weapon-laden battles to the bolts ... It's Robot Rumble time!

Photograph by Jon C R Bennett

Adobe Stock - eakgaraj

GETTING INVOLVED

Robot sports aren't just combat – there's something fun for everyone

Written by David Calkins

DAVID CALKINS is the founder of RoboGames and a former professor of robotics.

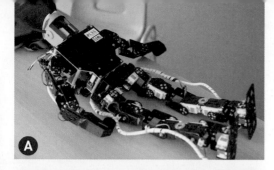

From 1997 to 2004, I competed around the world in a number of different robot categories, primarily sumo robots. I also worked a lot of shows (judges are always in demand) while teaching robotics at San Francisco State University. One thing I noticed in that time was not only the variety of robots out there, but also the variety of skills needed to build a robot.

And so I started RoboGames. The goal was to expose an audience at large to all the different sports (they'd come to watch combat, but stay for the soccer), and to cross-pollinate the contestants so that they'd become more familiar with other disciplines and possibilities.

Let's break down the categories of robot sports:

Ⓐ HUMANOIDS: The dream of all robot thinkers. They're usually only about 16 inches tall, but modern humanoid robots can walk, run, do cartwheels, backflips, and even play soccer and basketball.

Ⓑ SUMO: The oldest of robot sports, going back over 30 years. Basically, Rubik's Cube-sized robots that can think for themselves and autonomously find their opponent and push it out of the ring.

Ⓒ COMBAT: Just like NASCAR, it's exciting, big, and there are lots of crashes and explosions. While 100kg (220lbs) is the size of choice for the audience, there are six different weight classes as small as 150g (5.3oz) (see "Know Your Combat Robots!" on page 26).

Ⓓ SOCCER: Sometimes played with humanoids, sometimes with wheeled bots, this is the hardest of the robot sports. You have to be able to program the robot to do all sorts of moves and work with between 3 and 11 robots per team.

Ⓔ AUTONOMOUS CARS: Commercial autonomous cars still don't perform perfectly, and so we have autonomous robo-cars: line followers, NatCar (really fast line followers), and RoboMagellan, which is fully GPS driven.

ROBOT RUMBLE! Competition Primer

F HOCKEY: Three remote controlled robots to a side play hockey using a standard street puck.

G ART BOTS: Robots don't have to be limited to roles of speed and strength. What about beauty? We celebrate robots that make cocktails, paint, play musical instruments, or just look and act cool!

H OTHERS: If you can think it up, there's an event for it. Robots that balance like Segways, robots that can put out fires, solar robots, task-oriented robots … For a full list of events and rules, check out robogames.net/events.

At RoboGames, there are 54 categories in total. Many of the robots have a lot in common:

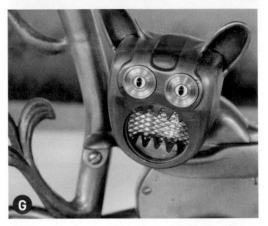

I MOTORS: Combat and sumo robots can get away with two motors or servos, while humanoid robots have from 18–30 of them. Motors range in cost from $1–$2,000 and servos are about $5–$200, depending on quality and abilities.

J ELECTRONIC SPEED CONTROLLERS (ESCs): These control the motors. Do you want to go fast or slow? The ESC determines that. ESCs cost from $30 on the tiny robot side all the way up to $1,000 for large-scale robots.

K BRAIN: The computer that runs everything. If you've played with an Arduino, you're already started! Humanoid robot brains are often specialized controllers that position each servo

like a human muscle or joint, as well as take all the input from the sensors to control how the robot reacts to its environment.

🅛 **BODY:** Here's where things get crazy. I've seen robots built from literal scrap metal to custom CNC'd titanium shells costing $20,000. Obviously if you're competing in large-scale combat you want something that's made of metal, while a 500g sumo robot is best made from plastic.

RECHARGEABLE BATTERIES: Robot batteries are usually lithium-based and come in either a C-cell shape or custom flatpacks. Good batteries are both expensive and dangerous (look up "lithium battery fires" on YouTube). You can get quality rechargeable packs from $50 on up, depending on how big your robot is. They're measured in both amp-hour charge (how long they'll last) as well as how fast they can be discharged. Beware: Draining a lithium battery *too* quickly is how fires start.

BATTERY CHARGER: Chargers are usually sold separately from their power supplies, so this is a two-fold purchase. Never buy a cheap charger — it'll ruin your batteries and you'll be out a lot of money.

WHERE TO BEGIN?

If you've never built a robot before, regardless of what you're into, my suggestion is to start with a kit. Why? Because you probably don't have all the tools needed to build something right away, and mistakes in robotics can be expensive.

Category-wise, there are three directions I recommend you take as a starter:

- **COMBAT:** Start small. I suggest beginning with a 3lb robot. You can get a kit for around $150. Add on two battery packs, a battery charger, and a transmitter and you're at about $400. (Also see "Combat Kitbots" on page 40.)
- **SUMO:** You can get a decent 500g (2lb) mini-sumo kit for around $100. If you want to learn to program robots around sensors, this is the best way to start. You'll learn about controlling speed, vision sensors, and programming for a real environment.

- **HUMANOID:** Because of the number of servos involved, humanoids start at around $1,000 and go up from there.
- **HOCKEY:** OK, a fourth option. If you're hell-bent on building your own robots, I'd suggest starting with a trio of hockey robots. Use battery packs and gearmotors from cheap drills, cut a frame from polycarbonate with a handsaw, pop in some scooter wheels, and then throw in an ESC and transmitter, and you've got yourself a hockey team.

The final factor in competing isn't so much what you build as where you live. Like many sports, you want to play against someone! To really compete in robot sports, you need to live near a bunch of people who also want to compete, and you need an arena (See "Cage Match" on page 54).

Either that or travel to a robot event. There are more of them popping up all over the world all the time, so really it's just a matter of Googling your city and "robot events" and working from there.

The best way to start is to just sign up for an event to give yourself a deadline. This will help to ensure that you don't leave your robot unfinished. If you're looking to compete, a good place to start is RobotCombatEvents.com which lists events all over the world.

Later this year, RoboGames will be returning and featuring all of the above mentioned events. You can find more information at robogames.net, including rules, photos, and help to get you started. Go out there and have fun!

THE FUTURE

Since I started in robot sports, I've always felt that robots were the future of sports. If you compare robots to the UFC, for example, robots heal much faster than humans. So while a human UFC fighter might only have 2–3 fights per year, robots could support weekly events, just like the NFL does — and even then, football players get injured all the time.

So a better comparison is NASCAR, but with the obvious difference that the robots are trying to kill each other, while in NASCAR they're just trying to outrun each other. But it's the crashes, fires, and explosions that make the highlight reel — which is what robot sports are all about! ⊘

KNOW YOUR COMBAT ROBOTS!

Written by Peter Garnache

A field guide to competition weight classes and weapons

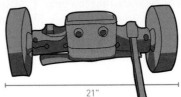

4.75"	7.25"	7.2"	21"	9.875"
FAIRYWEIGHT 150g	**ANTWEIGHT** 1lb	**BEETLEWEIGHT** 3lbs	**HOBBYWEIGHT** 12lbs	**DOGEWEIGHT** 15lbs
Micro Malice by Adrian "Bunny" Sauriol	Midas by Brian Latimer	Silent Spring by Jamison Go	Bobby by Jonathan Clark	Razors Edge by Owen Cokley

22"
FEATHERWEIGHT 30lbs
Emulsifier by Matthew Bores

34"
HEAVYWEIGHT 250lbs
Cobalt by John Mladenik

Caleb Kraft

PETER GARNACHE is a mechanical engineer, robot builder, and mentor to high school robotics teams. In the past 5 years, his 11 combat robots have 5 first place finishes and 13 top three finishes. repeat-robotics.com

The combat robotics community spreads across the entire world, with major competitions happening on almost every continent. Combat robots are divided into different weight classes based on the robot's overall weight, and within those classes an amazing number of weapon types are possible. In this guide we'll introduce you to all of the most important weight classes and weapons.

THREE FLAVORS OF COMBAT

In addition to weight, three different class modifiers are common in robot combat:

- **Full Combat** or **Standard** class is the least restricted ruleset, allowing kinetic energy weapons, wedges, and a variety of *control bots* (lifters, etc.). Most events run Full Combat classes.
- **Sportsman** classes add tip-speed limits to spinning weapons, drastically reducing their ability to do damage. They also require an active weapon on each robot, to stop passive wedges from dominating the competition. Sportsman classes are common at the 12lb and 30lb weight classes.
- **Plastic** classes, generally only found at the 1lb Antweight scale, require robots to be entirely fabricated from plastic parts (aside from electronics, motors, and fasteners). The Plastic Antweight is rapidly growing in popularity as an entry-level class due to the ease of 3D printing and the low damage rate, which make the robots extremely affordable compared to other weight classes. (Build your own on page 44.) You can learn more about combat rules online at sparc.tools.

ROBOT WEIGHT CLASSES

There have been many weight classes over the last 30 years, from 25g to 350lbs, but not all of them have active competitions today. Here's your guide to currently active combat robot weight classes.

FAIRYWEIGHT (U.K. ANTWEIGHT) — 150g limit

A good class for beginners. Many free designs can be downloaded from Thingiverse and printed on a 3D printer; components are cheap, and the lack of devastating weapon power makes Fairyweight bots cheap to repair. Unfortunately, motors and electronics take up more of the overall weight of a Fairy, as it is hard to find super small components, making custom designs challenging. Although Fairyweights are usually made from 3D-printed plastic, the most competitive are packed with titanium, steel, and carbon fiber to maximize strength and performance.

ANTWEIGHT (U.S.) — 1lb limit

Antweights are the best place to start out. Designs that are mainly 3D printed or fabricated with cheap materials can be competitive. Weight is not as much of a limitation as it is in Fairies. Common robot construction methods, such as directly bolting weapons and wheels to motors, can be used successfully with little modification. This class is generally less destructive than the Beetleweight class, so your robot is less likely to get completely destroyed. And because the electronics take up less of the entire weight, robots that are homemade with cutting boards and duct tape can easily survive and even win Antweight events.

BEETLEWEIGHT — 3lbs (U.S.) / 1.5kg (U.K.)

By far the most competitive insect weight class. Brushless motors at this scale provide insane power-to-weight ratios, thanks to the huge market for micro drones which has driven the improved power density and affordability of small motors. This makes Beetleweight robots incredibly powerful and dangerous. While they can be made by hand and with 3D-printed parts, the most successful Beetleweights feature a multitude of custom machined parts in exotic materials. Solid engineering and design are a must for this class, as robots regularly get ripped in half.

While you might expect bots in this class to be mostly metal, a surprising amount of engineering plastics are used, giving superior impact protection and strength while minimizing weight and cost. Additionally, 3D-printed materials such as nylon and TPU are heavily utilized, making fabrication much easier than with larger bots.

HOBBYWEIGHT — 12lbs

Hobbyweight bots are the smallest non-insect bot. This weight class allows for more creativity in robot design. The larger footprint provides space for more complex mechanisms, while the larger weight budget allows for more off-the-shelf components to be used, and the motors and electronics take up a smaller proportion, leaving more weight for frame and weapons. The weapons on these robots store incredible amounts of energy, so they must be designed more robustly than bots in smaller classes. Plastics are no longer the go-to for frame material; Hobbyweights start to use aluminum and steel frames in order to survive the devastating weapons that they face. While these bots are four times as large as Beetleweights, they're nowhere near as power dense. No micro drone motors here — Hobbyweight bots work at a larger scale, with motors that are generally used in large RC cars and electric skateboards.

DOGEWEIGHT — 15lbs

Dogeweight robots are essentially overweight Hobbyweights. Most competitions that host larger bots host 12lb bots, but this 15lb class has found life in an educational setting. Leagues such as NRL and Xtreme BOTS host high school and college level competitions where engineering students can practically apply what they're learning to fighting robots. As with Hobbyweights, these robots are incredibly dangerous and can do amazing amounts of damage.

FEATHERWEIGHT — 30lbs

This weight class is very popular for experienced builders, as a single builder with experience can design and fabricate a Featherweight. There are several competitions for these robots annually in the U.S. and worldwide. Featherweight robots are generally less weapon oriented than smaller robots and focus more on reliability and defensive armor. Good engineering becomes more important when fighting at the featherweight scale because parts need to be relatively lighter than smaller scale robots while providing the same performance.

LARGER WEIGHT CLASSES

Competitions for robots above the featherweight class are few and far between. Lightweights and Middleweights are practically extinct, and Heavyweight events have been limited to well-funded TV shows, or small sportsman events. The cost of building these robots grows exponentially between weight classes, so it takes a well-sponsored team with experience to compete at this scale. The best-known combat robot event is *BattleBots*, which is televised on Discovery and has near 1 million viewers each episode. Outside of *BattleBots*, various *Robot Wars* TV series from the U.K. are seen around the world, and several heavyweight competitions have been filmed in China including *FMB (Fighting My Bots)*, *Clash Bots*, and *KOB (King of Bots)*.

LARGER WEIGHT CLASSES

COMPETITION CLASS	WEIGHT
LIGHTWEIGHT	60LBS
MIDDLEWEIGHT	120LBS
INTERNATIONAL HEAVYWEIGHT	110KG
BATTLEBOTS HEAVYWEIGHT	250LBS

SPINNING WEAPON TYPES

Here's how they work, and their advantages and disadvantages, especially in the smaller classes.

VERTICAL SPINNER

These weapons store energy in a spinning mass that rotates in a vertical plane, generally sticking out the front of the robot (Figure Ⓐ). They spin so that the impact point of the weapon is moving upward, effectively delivering an "uppercut" to the opponent. Vertical spinners are very effective fighters, as the reaction force from the uppercut pushes the bot downward into the floor, keeping them planted.

The key to the vertical spinner is the ability to get under other robots. When an opponent rides up a vertical spinner's wedge or forks, the spinning blade can impact the bottom edge of the opponent's chassis, rather than grinding against

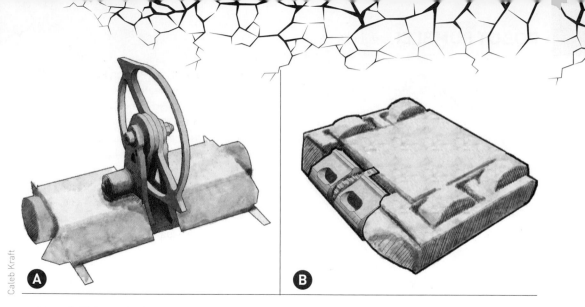

a side wall. This results in significantly better energy transfer, meaning a harder hit and more damage.

Vertical spinners generally spin either a bar or a disk weapon, and commonly switch between the two depending on the opponent they're facing. Disks are more popular because they store more energy for a given mass, but they're vulnerable to horizontal spinners. Doubling the spinning speed of a weapon results in four times the kinetic energy, so vertical spinners will often deploy asymmetrical, single-tooth weapons in order to maximize their "bite." This allows them to save weight while increasing the energy stored in the weapon.

One problem vertical spinners can suffer from is gyroscopic forces on the robot due to the high momentum of the weapon. Like a bike wheel, the spinning weapon acts as a gyroscope and resists the robot's turning motion. Turn too fast, and one end of the robot "floats" off the ground. This effect can be used to self-right some robots but it can also flip a robot upside down. Because of this effect vertical spinners need to limit their turning speed, or turn in large arcs. These handicaps can give other robots an advantage.

DRUM SPINNERS

These weapons also spin vertically but they're wider, with smaller diameters, and can spin at much higher speeds. Drum spinner robots are mostly driven with two-wheel drive and have smaller and more compact frames than any other weapon type (Figure **B**), but their geometry

is especially prone to float when turning.

Because the weapon on a drum spinner protrudes past the frame of the robot, their ground clearance is not as important as other vertical spinners. Most robots get away with a small piece of metal pointed toward the ground to stop other robot's wedges from getting under, but even that's not absolutely necessary to be competitive.

There are several popular ways to make the drum weapon. The first is to fabricate a single-piece drum out of a hardened steel. This is difficult without highly specialized tools. An easier way is to mount hardened steel "teeth" to a drum using bolts, but these weapons have to be manually balanced which takes considerable time. The easiest way is by layering hardened steel disks with spacers made from a softer material, such as aluminum. This simulates the large impact face of a unibody drum while also storing a large amount of energy in the rotating mass of the weapon.

BEATER BAR SPINNERS

Beater bars are a weapon type that sits between verticals and drums. They span most of the width of the bot, like a drum, but they can store more energy with a larger diameter, like a vertical spinner. The general idea behind a beater bar is to maximize the moment of inertia (MOI) of the weapon while keeping the weight low. This is achieved by removing most of the material on the inside of the weapon, and keeping the outermost material, as mass further from the axis of

C

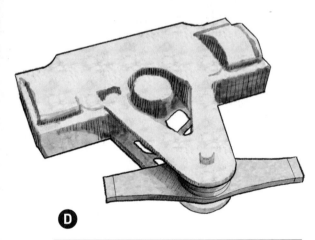

D

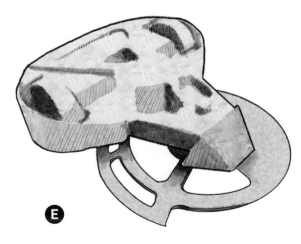

E

rotation has more of an effect on MOI (Figure **C**).

The most effective beater bars are solid steel, custom machined using CNC or wire EDM processes, but there are good low-cost alternatives. These budget weapons are cut out of a flat piece of aluminum in an outline of a rectangle, with a hollow interior; bolts are mounted to the top and bottom edges to act as a cheap hardened impacting surface. These can be cut with only a few operations and are nearly as effective as more expensive weapons at a fraction of the price.

HORIZONTAL SPINNERS

These weapons spin in the horizontal plane and are generally much larger in diameter than vertical spinners, but they spin at slower speeds. Most horizontal spinners can be split up into one of three categories:

Horizontal bar spinners or **midcutters** have their weapon mounted off the ground, between the bottom and top of their chassis (Figure **D**). Horizontal spinners generally have bar weapons, as they are vulnerable to vertical spinners. Using a disk weapon this high off the ground will give vertical spinners great engagement into the weapon, making bending more likely. With a bar, there's less surface for a vertical to hit, and the weapon can be made stronger, as it does not have to span as much area. Horizontal bar spinners are vulnerable to wedges, as their weapons struggle to get good bite on hardened surfaces and get deflected upward. These robots are also very prone to self-inflicted damage, as their weapon can be forced into a floor or wall, causing the robot to ricochet around the arena.

Undercutter weapons are mounted under the chassis, as close to the ground as possible (Figure **E**). Undercutter robots attack low on their opponents with the goal of bypassing armor and directly hitting important components such as wheels. Because they spin so low to the ground, other robots have trouble getting under them without taking significant damage, and they're less vulnerable to vertical spinners. Undercutters can use either disks or bars for weapons, but disks are more widely used as they store energy more efficiently.

Overhead bar spinners can spin weapons

with much larger diameters than undercutters or midcutters, because they spin over the top of the entire robot chassis (Figure **F**). These robots are very fun to design and build and can be unbelievably destructive due to their high energy storage, but they have several inherent flaws. They can't drive while inverted, as the weapon lies between the wheels and the ground. Because they store more energy, they take the greatest strain on the chassis and weapon system whenever they land a hit. They're also extremely vulnerable to vertical spinners as their weapons are mounted so high and cannot be supported on both top and bottom.

FULL BODY SPINNERS

Full body spinners have a spinning weapon that protrudes outside the robot's drive chassis and surrounds the entire robot. They function as both weapon and armor, and usually weigh more than any other weapon type. Full body spinners are split up into three types:

Ring spinners spin a ring weapon around the robot's frame (Figure **G**). These rings are supported at the outside of the robot and do not cover the top. Therefore the robot can use wheels that are taller than the frame and stick out the top, which allows them to drive while inverted. While being invertible is a huge advantage, this weapon system is inherently fragile. Because it spins around the body of the robot (with several contact points to the frame) the weapon will fail with any deformation. This makes ring spinners one of the most complex weapon types to design successfully.

Shell spinners have a weapon that covers their top and sides, protecting the entire robot (Figure **H**). They can't drive when inverted but have a much simpler weapon support compared to ring spinners. Shells generally have a protrusion on top that makes them unstable when inverted, allowing them to right themselves. These weapons are much more robust to damage, as they generally spin about a shaft in the center of the robot and deformation of the outer shell does not necessarily mean it will be disabled.

One problem that plagues both ring and shell spinners: because their weapons have large MOIs

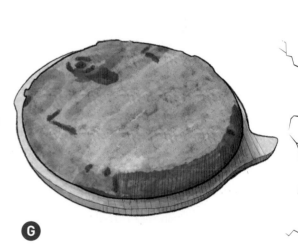

F

G

H

Caleb Kraft

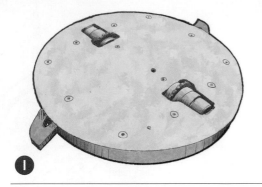

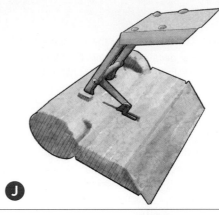

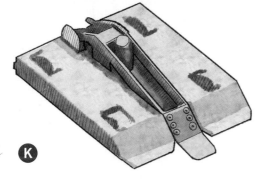

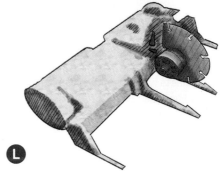

(due to the large radius and weight), they're prone to losing drive traction when the weapon first starts spinning, which starts the chassis rotating in the opposite direction. This causes the driver to completely lose control of the bot.

Meltybrain spinners are a special type of robot that uses its entire body as a weapon and its drivetrain to power it (Figure **I**). Meltybrain robots have very high-speed drivetrains that essentially spin them in place. They can then use sensors to modulate the speed of the motors to achieve translational movement. Because the full robot weight is in the weapon, these bots hit extremely hard. Their entire chassis must withstand thousands of RPMs, so they are generally very tough. While meltybrains are extremely damaging, they are very technically complex and require more computing than any other robot type.

NON-SPINNER WEAPONS
LIFTER ROBOTS
Robots that have a mechanism that does not store energy and is used to lift up and flip over other robots are called *lifters* (Figure **J**). These weapons move slower than *flippers* (see below),

and can be moved to any point along their travel.

Most insect-weight lifters use a high-power servomotor or gearmotor to power their lifting arms through a simple mechanism such as a lever or a four-bar linkage. Lifters are the safest bots that have active weapons, as they do not store any energy in their weapon system. (You can build one on page 44.) Some lifter robots also have integrated grabbing mechanisms that can hold their opponent as they lift them off the ground. Lifter robots can be very competitive in arenas that feature some sort of pit or "arena out" feature, as they can corral their opponent into the out-of-bounds area, winning the match. Lifters are strong against robots that are not invertible, as a single flip will render their opponent immobile, but they generally struggle to cause enough damage to score knockouts.

FLIPPER ROBOTS
Unlike lifter robots, flipper robots have a weapon that uses stored kinetic energy to lift or flip their opponents (Figure **K**). Flippers feature complex mechanical systems that make reliability a challenge: springs or pneumatics for insect-weight flippers, spinning flywheels or

hydraulics for the heavier classes. Some events ban pneumatic weapons due to safety concerns, which makes flipper robots even harder to successfully compete with. Flipper robots work best in an arena with an "arena out" feature but can be powerful enough to damage robots and achieve knockout victories in a standard arena.

OVERHEAD SAW ROBOTS

These robots have a spinning weapon at the end of an arm that can be independently articulated. Most have an actual saw blade (Figure **L**) that they use to cut through other robots, taking advantage of the lack of top armor. (Others have a weapon called a *hammer saw* that's no different from a regular spinning weapon as it imparts kinetic energy through an impact.) Overhead saw bots usually have a very open front end with prongs protruding, as they need to control other robots and hold them still for the saw blade to be effective. These robots require a robust design and also a talented driver, as the articulated weapon is more demanding to operate and they usually need to pin their opponent in place to do meaningful damage.

HAMMER ROBOTS

Hammer robots and their cousins, axe robots, have a weapon that swings a mass over the top of the robot in a limited arc to impart kinetic energy into their opponent (Figure **M**). At the insect weight classes, these robots can be powered with either a high-speed servo or a motor. Like overhead saws, these robots take advantage of the light top armor on their opponents. There are two main problems with insect-weight hammer bots. First, their weapon must be triggered slightly before the opponent is in front of them because the hammer takes time to complete its swing. Insect bots move fast relative to their larger counterparts, which exacerbates this problem. Second, they struggle to store comparable energy to spinners as their weapon only has up to 180 degrees of motion to accelerate before impact.

CRUSHER ROBOTS

Crusher robots have an overhead arm that pierces through the top of their opponents,

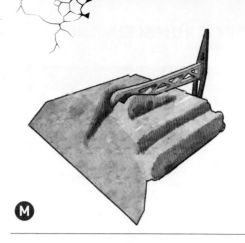

M

N

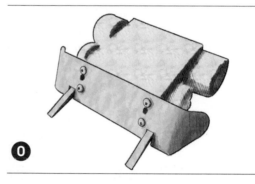

O

damaging the internals (Figure **N**). At insect weight classes, crushers are designed to convert fast rotational movement from a motor into slow, powerful linear movement at the tip of their claw. The fatal flaw of crushers is that their actuators either are too slow to easily grab their opponents, or lack the force needed to pierce armor. Crusher robots are mechanically complex and require extremely precise driving.

WEDGE ROBOTS

Passive wedge bots (Figure **O**) don't have an active weapon but they're very durable, as they rely on pushing their opponent around until they break. These robots are the simplest of all — and very competitive! ●

EXPECT TO BE DESTROYED

Austin McChord is the CEO of Casana, and the founder and former CEO of Datto, Connecticut's first "unicorn" startup. He is a graduate and trustee of the Rochester Institute of Technology.

I started **Norwalk Havoc Robot League** to bring the professional experience to **all** combat bot competitors

Written by Austin McChord

My first experience with combat robotics ended exactly as I expected: in complete destruction.

It was 2016. A friend of mine, Leanne Cushing, competes on *BattleBots*, and she suggested I join her for a combat robotics competition in Massachusetts. Not knowing much of anything about robots in general, or combat robots specifically, I built two bots and was unceremoniously crushed in the competition. It was totally and completely awesome.

I knew from that moment that I enjoyed the sport. The challenge and the fact that you are competing with something you build and design is just really unique. As a builder, however, I had this feeling that there was something missing. The event I'd been at felt like it was poorly paced, making for a long day without enough action to keep competitors engaged.

THE CAGE IS THE STAGE

Fast-forward to 2018. I had just sold Datto, the company I founded in 2007, and trust me when I say I'm not the kind of guy who can sit still for long — I was itching to build something. Anything. I knew I had the operational background of bringing order to chaos in running a company, so I decided to try to put on a combat robot event of my own. I bought an 8'×8' cage from a group associated with the New York City Maker Faire, and posted a message to a few corners of the internet inviting people to come fight robots with me at an office building that I own in Norwalk, Connecticut.

To my surprise, it actually worked. In September of 2018, 17 people showed up to stand around a little cage on the third floor of an office building to fight robots. That was our first event. We made a "house bot" out of a cinder block, and livestreamed fights to YouTube. I was the producer, announcer, and video switcher, and spent the day just trying to keep up. It was a blast.

We held a few more events in 2019. In the beginning, the competitors were mostly from the northeastern U.S., but soon people started coming from farther away. We started seeing 30 or 40 robots show up at each event. At each successive event, the flow of matches went faster and faster, because I wanted to make them really

Original home at 50 Day Street, 2018.

I KNEW I HAD THE OPERATIONAL BACKGROUND OF BRINGING ORDER TO CHAOS IN RUNNING A COMPANY, SO I DECIDED TO TRY TO PUT ON A COMBAT ROBOT EVENT OF MY OWN.

Original control room and cage in 2018.

New control room in action.

watchable via livestream. We fought them all in just that single cage through the end of 2019.

I had run a startup before, so I knew to focus on achievable, constant, incremental improvements. After every event, the question was "How do we improve the production quality a little bit? How do we run this event a little bit more efficiently?"

It was slow and steady: First, we brought on new announcers. Then we brought someone in to help switch the video feeds. And so on and so forth. For the first event, we wrote the bracket out on a whiteboard. Now, the brackets and match flow and everything are deeply computerized and automated. It's been a process, but we haven't stopped focusing on identifying and making high-impact improvements after every tournament. Sometimes it's one step forward, two steps back — but we are never complacent and always working to improve the experience across the board.

THE COVID HIT

While we prepared for our first event in 2020, the pandemic happened, and we had to push pause on hosting tournaments. Covid was, however, like an accelerant on all things, so it's also when things got interesting for our little league. I ended up with too much time on my hands (which, as noted above, leads me to do crazy things), so, when I had the opportunity to buy a 67,000-square-foot building right down the block from the office building I already owned, I went for it.

Owning our buildings gives us a lot of advantages. We've been able to install a lot of safety equipment that, for instance, allows our cages to run at a negative pressure, which means that we can have things like fire (always a spectacle); in another environment this wouldn't be safe. Our cages are made of steel and have two thick layers of Lexan, which is similar to

I ENDED UP WITH TOO MUCH TIME ON MY HANDS (WHICH, AS NOTED ABOVE, LEADS ME TO DO CRAZY THINGS), SO, WHEN I HAD THE OPPORTUNITY TO BUY A 67,000 SQ. FT. BUILDING RIGHT DOWN THE BLOCK FROM THE OFFICE BUILDING I ALREADY OWNED, I WENT FOR IT.

Outfitting the new location at 165 Water Street.

The team pits.

WE'VE BUILT A STRONG COMMUNITY OVER THE YEARS, AND WE FEEL STRONGLY ABOUT PUTTING OUR BUILDERS FIRST, BECAUSE WE'D BE NOWHERE WITHOUT THEM.

bulletproof glass. We can have batteries explode; even rocket motors can be used in a safe way. No one in the audience is at risk, none of the drivers are at risk. All this safety means more options for builders to get creative with their designs.

With safe procedures and more space, I knew we could host bigger, badder robot fights. That would mean more cages, bleachers for fans, and a lot more media gear. We went up to four weight classes: 3lb, 12lb, 12lb Sportsman, and 30lb.

The second half of 2020 was spent outfitting that building to be ready to have robot events again, starting in February 2021. We were crazy enough to say "come and bring your robot to fight" — and people did, in droves. In February we had 49 robots, and set up the pits, where builders do their repairs, outside in the parking lot. At our next, in March, we had 68; then 83 in May and 111 in July.

It was easy to see we were on to something. We were seeing more people, coming from farther and farther away, and as Covid receded and travel got a little easier, we just kept getting bigger and bigger.

READY TO GO BIG

That was the point when I realized that Norwalk Havoc was becoming more than a hobby. The community was starting to see us as the future of

this niche but growing sport. We were getting big, and really competitive, and I knew we needed a dedicated team in order to exist long-term.

In the summer of 2021, I started looking for someone who could run the league full time, and help monetize and scale it beyond the walls of our warehouse. I was introduced to Kelly Biderman, who had years of experience in media working at places like *The Wall Street Journal* and Katie Couric Media, but no background in robotics. I laid out the opportunity — noting that she would be *absolutely crazy* if she took this job — and perhaps she didn't believe me, but she came on board as CEO of our new company, Havoc Robotics, in the fall.

Kelly's "outsider" perspective has been really valuable so far; she s brought in a professional production team and content strategist who have leveled up our livestream to resemble a broadcast-quality program, and helped us increase our digital audience by several thousand by marketing our massive content library more efficiently. She has big ideas and ambitions around sponsorships and partnerships with brands that share our values. Those values: making STEM fun; encouraging innovation and creativity while pushing limits, safely; and celebrating and supporting builders and makers.

We've built a strong community over the years,

Match announcers.

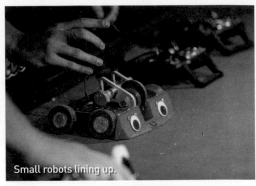
Small robots lining up.

and we feel strongly about putting our builders first, because we'd be nowhere without them. That's part of why we're exploring sponsorship as the avenue to monetize the tournaments right now — because the cost to compete with us is $0. Robot fighting is an expensive sport, so we make a point to limit the cost that we pass on to our competitors, and instead find ways to support our league, and eventually our builders, with the help of like-minded brands.

WHERE WE'RE GOING

As of March 2022, we have five cages in operation at each of our seven annual tournaments, and we rotate fights between them as quickly as we can. Because so much of this is geared around the livestream, and I'm a nerd, we have over 70 cameras to follow the action from every angle. We've expanded our team to include replay operators, a professional TV graphics developer, and a pit reporter. Our March 2022 event was our largest event ever, with 112 robots, over 400 spectators, and more than 25,000 views of our livestream on YouTube. For spectators in person, there are bleachers surrounding the cages and there are television screens everywhere so no matter where you are in the arena, you have a great view.

After events, we publish and package individual fight videos, which are a more accessible way for new fans to discover the sport than a 12-hour livestream. What we're really working on now is figuring out how we attract those adjacent audiences who are maybe fans of *BattleBots*, or similar things like esports and gaming, and bring them into the fold by just showing really cool videos of robots blowing up and destroying bulletproof glass.

This sport has serious competitors in the same way that mainstream and esports do, and now it's really about growing the following. I expect to find that some of our robot fighters end up being meaningful influencers in their own right, and we're eager to support that. Builders have

Match sparks a-flyin'.

Jon C R Bennett / JCRBPhoto

Something's getting shredded!

The crowd approves.

amazing stories and journeys that brought them to this sport. They're creating a little brand for each one of these robots, and you can see people getting excited about them, and that's just really cool to witness.

We truly believe our potential audience is virtually untapped, and that we can bring this sport from niche to mainstream. It doesn't matter who you ask — if you show someone clips from our events, they can't stop watching. It is really fun to compete, whether you're a first-timer or a pro. It is a fantastic way to gather a whole bunch of innovative, creative, STEM-minded folks together and see how they are tested.

We're a little nerdy, a little quirky — we don't take ourselves too seriously, which makes us welcoming for people of all experience levels and backgrounds. We award winners with cash prizes, but as a nod to our irreverent beginnings, they're also handed a miniature dumpster as a trophy. We want to keep that bit of weirdness in place because we're all a little weird, and owning

and celebrating that has helped us build our community into what it is today.

We have everyone from rocket scientists and professional roboticists to 12-year-old tinkerers and groups of moms in their 40s. They are incredibly supportive, and are constantly helping each other in the pits or on Discord, because they want to compete against the best robots out there, and still have fun doing it. It is truly a sport for anyone, and we have a laser focus on making sure our league is builder-first and accessible for all.

We hope that by the end of this year, we'll have the tournaments running in a predictable, efficient way, because we have our sights set on growth beyond Norwalk, and even beyond combat robotics events. You can really feel the potential buzzing around Norwalk Havoc, and for us, this is just the beginning. ◗

 Find Norwalk Havoc Robot League online at nhrl.io and youtube.com/c/NorwalkHavoc.

WE'RE A LITTLE NERDY, A LITTLE QUIRKY — WE DON'T TAKE OURSELVES TOO SERIOUSLY, WHICH MAKES US WELCOMING FOR PEOPLE OF ALL EXPERIENCE LEVELS AND BACKGROUNDS.

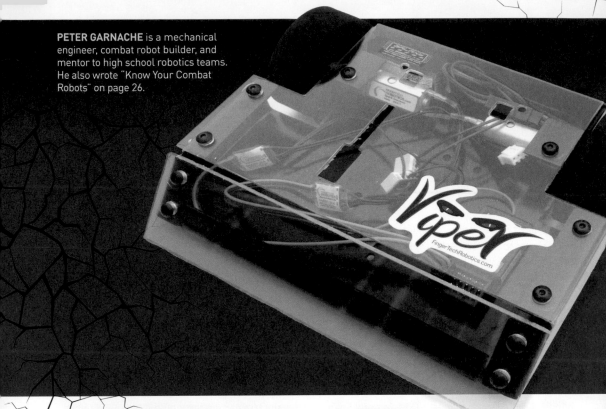

PETER GARNACHE is a mechanical engineer, combat robot builder, and mentor to high school robotics teams. He also wrote "Know Your Combat Robots" on page 26.

COMBAT KITBOTS

So you want to fight robots? Try these battle-hardened kits for building antweight (1lb) combat bots Written by Peter Garnache

Ever since I was a kid, I've wanted to fight robots. I saw *BattleBots* on TV and knew that one day I would have a robot of my own to compete. It wasn't until I reached college, pursuing a degree in engineering, that I got the chance to build my own robot. Most people who watch *BattleBots* don't know that there are many weight classes outside of 250lb heavyweights. In fact, there are hundreds of competitions for 1lb and 3lb robots around the USA and the world.

But getting started in combat robotics with a completely custom design can be a big hurdle

to get over. There are thousands of choices for motors, batteries, and speed controllers, and having to decide everything on your own can be overwhelming.

Thankfully, several vendors have designed kit robots that you can purchase, assemble, and then really compete with. These builders have done all the hard work sourcing motors, programming speed controllers, designing custom weapons, and fabricating most of the parts. This destroys any barriers to entering your own bot into battle. If you think fighting robots sounds like fun, then I highly recommend starting with a kit robot. You'll learn more than you'd expect from your first event, and most of the kits are pretty competitive for their cost.

Most kits come with everything you need to compete, other than an RC transmitter and a lithium polymer battery charger. These two components, sold separately, can be shared between many robots and are things that you'll keep using when you move forward and start building your own bots.

I've now designed and built 11 combat robots and fought over 100 fights. I've had a chance to fight against just about every kit on the market, and the few I haven't fought, I've watched fight at several events. Here's my advice on getting started with antweight (1lb) kitbots.

BEST FIRST KIT: THE VIPER

While you might be excited to get a kit robot with a big spinning weapon, I caution you against a kinetic weapon for your first robot. These bots are not toys, and can easily release more energy than a handgun. When testing the spinners without proper safety precautions and a test box, it's really easy to end up in the emergency room. I recommend starting with a wedge or lifter kit robot and once you get some experience, moving to something more dangerous.

The first kit that anyone venturing into combat robotics should get is the **Viper** from **FingerTech Robotics** (fingertechrobotics.com) (see "The Roboteers" on page 50). This antweight bot features a bent sheet metal frame, two drive motors attached to foam wheels, and a polycarbonate wedge. It truly is the simplest kit around, meant for someone who hasn't touched a fighting robot. There are really good instructions at their website, and the $156 price point makes it the cheapest kit on my list by a long shot. While you might think a simple kit like the Viper can't win matches, they're actually very hard to kill. Upgrading the front wedge from polycarbonate to steel or titanium allows them to take heavy spinner hits, and wheel guards can easily be added to protect their drivetrain. Overall a great investment and the best starter kit on the market.

Once you've finished your first competitions with the Viper kit, you don't have to throw it away

to move on to something new. FingerTech offers three upgrade options for the Viper:

- The **lifter module** is the cheapest of the three, adding a servo-operated lifting arm made of polycarbonate and steel. This attachment gives you an active weapon that is not dangerous, and can be tested outside of a test box. It gives your Viper the ability to flip other bots over to cause knockout wins!

- The overhead **horizontal spinner module** is by far the most common of the Viper weapon kits, as it sports a massive weapon and can create exciting fights. Be careful fighting it against vertical spinners, as big hits can detach the blade or motor.

- Finally, the **vertical spinner module** comes with a clamping drum with AR400 steel teeth. This weapon hits hard and is the most durable of the three modules. Be careful testing it on your own: the spinner is dangerous and requires proper safety precautions (see page 54).

ROBOT RUMBLE! Combat Robot Kits

FingerTech also sells an antweight-scale version of their popular aluminum **beater bar** weapon system. This is not a full robot kit; rather, they sell the weapon and the weapon electronics as a separate kit. This is great for a second or third robot, but not great for someone who hasn't made things before, as you'll have to design the chassis and drivetrain yourself. I've seen a few fights with these weapons, and if designed well they offer a great balance of weapon power and robot reliability.

DURABLE AND DESTRUCTIVE

Another good vendor for antweight kits is **Battle Robot Kit** (battlerobotkit.com). They sell four different kits that all sport powerful drivetrains and extremely durable single-piece chassis made of UHMW (ultra high molecular weight) polyethylene. All of these bots start at a base price, with a $100 "competition upgrade" that swaps in premium parts and accessories. Overall these kits are extremely solid and consistently have great performances at the events I go to. Each is well worth their price tag, as they are some of the most durable kit bots on the market.

- **Murder Hornet** is a forked wedge bot with 4-motor drive, and AR500 hardened steel forks up front; it's safest for beginners.
- **Taserface** is a very popular vertical spinner that is easy to assemble and hits hard. It has AR500 forks that can easily get under its opponents as well.
- **Lobotomy** features a massive undercutter that I have seen rip opponents in half. It is truly a terrifying bot to face.

- **Hellraiser** is a drum bot that uses bolt teeth to rip parts off its opponents. While it might not have the same impressive list of event wins that the other bots have, it packs a punch and is incredibly destructive.

VIRTUALLY UNKILLABLE

If you're looking for a safe bot to practice driving with that won't take much damage, look no further than the **Candy Wasp** wedge from **BotKits.com**. This is a perfect bot for the first-time competitor who's worried about spending money on a kit and getting it ripped apart. The Candy Wasp is made from two pieces of billet aluminum with polycarbonate top and bottom plates, and a titanium wedge with all aluminum mounting components. It features 4 powerful 16mm motors with foam wheels that help absorb shock and protect the internals.

Just like its bigger brother, the D2 (Dozer Two) beetleweight kit, the Candy Wasp is virtually unkillable and, with a good driver, a contender for any event it shows its face at. And it comes in 7 different colors of anodized aluminum, giving it more visual flexibility than any other kit available. Whenever I'm up against one of these bots I know it'll go the distance. They are extremely tough to kill no matter what weapon type they are facing.

ADVANCED WEAPONRY

Do you like chaos and destruction? The **Shock!** undercutter kit from **Absolute Chaos Robotics** has the largest weapon of any of these antweight kits. I've seen these bots hit their opponent and send both themselves and the other bot into opposing arena walls. Luckily I haven't had to face one of these kits at an event, because I believe they could've completely destroyed just about any of my bots. If you want a robot that does maximum damage and puts on a show while doing it, this is the kit for you. But because

Murder Hornet

Taserface

Lobotomy

Hellraiser

FingerTech Robotics, Brandon Kittredge, Ryan Clingman, Peter A Smith

42 makezine.com

its weapon is absurdly massive, it's not suitable for beginners. The kit requires soldering and advanced assembly, making prior experience a must. It's also extremely dangerous if operated outside of a proper test arena, and I would never recommend doing so. You should always make sure you're safe when testing any combat robot, but these kits are the most chaotic bots I've seen, so I would take extra safety precautions.

The final kitbot I'll mention is the **Saifu** drum robot from **Kitbots**. Like the Candy Wasp and the Shock! bot, this kit's design was based on the success of a 3lb beetleweight robot (Weta-God of Ugly Things). It features a machined aluminum drum with an integrated motor, eliminating belts as a possible failure mode. The aluminum weapon has hardened steel bolts as teeth, which are replaceable, making the whole bot more serviceable. The chassis is UHMW, with wheel guards and carbon fiber top and bottom plates. While this bot is the most expensive kit on the list by a large margin, it makes up for it in durability and competitiveness. Saifu kits are contenders to win any event they enter, and can stand up to abuse from even the toughest-hitting bots. Their weapons just never seem to die. This bot has been iterated over time, and the current version 2.1 is the result of years of evolution and tuning. As with the other bots with active weapons, this bot is for advanced and experienced users only. ◉

Candy Wasp

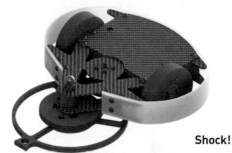

Shock!

Saifu

KITBOT NAME	WEAPON	PRICE	CLASS	DIFFICULTY	VENDOR
FINGERTECH VIPER	Wedge	$156	Antweight	Beginner	FingerTech Robotics
VIPER LIFTER	Lifter	$204	Antweight	Beginner	FingerTech Robotics
VIPER VERTICAL	Vertical spinner	$218	Antweight	Intermediate	FingerTech Robotics
VIPER HORIZONTAL	Overhead spinning bar	$227	Antweight	Intermediate	FingerTech Robotics
FINGERTECH ANT BEATER	Beater bar	$82	Antweight	Advanced	FingerTech Robotics
LOBOTOMY	Undercutter	$299	Antweight	Intermediate	Battle Robot Kit
TASERFACE	Vertical spinner	$299	Antweight	Intermediate	Battle Robot Kit
HELLRAISER	Drum	$349	Antweight	Intermediate	Battle Robot Kit
MURDER HORNET	Forks	$299	Antweight	Beginner	Battle Robot Kit
CANDY WASP	Wedge	$299	Antweight	Beginner	BotKits.com
SHOCK!	Undercutter	$299	Antweight	Advanced	Absolute Chaos Robotics
SAIFU	Drum	$676	Antweight	Advanced	Kitbots

BUILD YOUR FIRST COMBAT ROBOT!

Fast and easy to build, this 3D printed "antweight" is great for learning robot construction and battle skills

Written and photographed by Brandon Bennett Young

TIME REQUIRED:
3–4 Hours + 1–2 Days Print Time

DIFFICULTY: Easy–Intermediate

COST: $300–$400

MATERIALS:

Most of Kerfuffle's parts come from FingerTech Robotics (fingertechrobotics.com) which provides all the components you need to make 1lb and 3lb combat robots:

» **Gearmotors, Silver Spark 16mm, 22.2:1 gear ratio (2)**
» **Electronic speed controllers, tinyESC (2)**
» **LiPo battery pack, Galaxy 3S (11.1V) 300mAh**
» **Voltage regulator, 9V 4.5A**
» **Foam wheels, 2.00"×0.75"**
» **Lite Hubs (1 pair)**
» **R/C transmitter, T6A**
» **R/C receiver, TR6A**
» **Mini power switch**
» **JST connectors, 2-pin (1 pair)**

» **Servomotor, ANNIMOS 20kg metal gear** Amazon #B076CNKQX4. If you substitute the HXT 12kg from FingerTech — it has less power but still plenty for 1lb robots — you'll also need to buy the FingerTech metal servo arm.
» **Plastite screws: #4×³⁄₈" long (5) and #6×½" long (4)** from a local hardware store
» **Machine screws, M3×15mm long (2)**
» **Heat-shrink tubing** from hardware store or Amazon
» **3D-printed robot parts: chassis, lifting arm, top plate, and wheel hubs** Download the free 3D files from grabcad.com/library/kerfuffle-1lb-plastic-ant-combat-robot-1.

TOOLS:

» **3D printer with PLA+ or PLA filament**
» **Phillips head screwdriver**
» **Soldering iron and lead-free solder**
» **Hot glue gun**
» **Heat gun**
» **Hex wrench, 3/32"** to operate the mini power switch
» **Hex wrench, 0.050"** to tighten Lite Hubs setscrews

BRANDON BENNETT YOUNG has been building and fighting combat robots for the past decade in weight classes ranging from 150g to 250lbs, most famously Big Dill and Mammoth from the TV show *BattleBots*.

Want to build your own battle robot? Today's the day. Kerfuffle is a mini bot designed to inflict mechanical damage to other machines in caged combat. If you've ever seen the show *BattleBots* then you already have a very solid idea of how these robots operate.

Kerfuffle is a 1lb robot in the plastic antweight class, meaning it's not made with any of the high-grade metals or plastics, such as steel or nylon, that you may see being used on the heavyweight robots on TV. Kerfuffle is designed as an entry-level robot using inexpensive, 3D-printable materials that allow for many more people to take their first steps into the world of combat robots. For weaponry it relies on a wedge shape to get under opponents and a lifter arm to flip them over, meaning it's safe for beginners to practice — and even fight other Kerfuffles — without need of a protective arena.

I designed the machine originally to fight in my school's competition and it proved to be very effective. Since the first version in 2019, Kerfuffle has been tuned to become more competitive, leading to the successful Version 2 that you'll build in this guide.

Version 1 of Kerfuffle, 2019.

PRINTING THE PARTS

The first and most critical tool you'll need for this build is a 3D printer. The Ender 3 from Creality is one of the most popular and least expensive printers on the market with a price tag around $200. Other quality printers like the Prusa i3 MK3S are also great options, especially for higher quality materials like nylon, but they cost closer to $1,000. For our Kerfuffle, an Ender 3 is more than sufficient since we will be using PLA or PLA+ filament which are much less expensive and do not require a high level of tuning to print. These machines can be purchased online or in-person at stores such as Micro Center. Or you can send the 3D files out for printing by a service such as Shapeways.

Alongside the printer, be sure to pick up a roll of PLA+ filament in whichever color you prefer. Brands like Duramic3D and Inland have produced materials that have held up well in long-term use. Regular PLA can work too but I recommend PLA+ because it's tougher.

For print settings, I recommend roughly 4 walls and 50% infill. These settings can vary as you have more experience both printing items and fighting robots over time, but these will serve as a nice starting point.

BUILD YOUR KERFUFFLE COMBAT ROBOT

Once you've printed your parts and received the components, we can get to the fun part: building!

1. SOLDER THE ELECTRONICS

FingerTech's diagram (Figure **A**) shows a typical wiring setup with a gearmotor and servo. But we'll solder our connections, and also add a JST battery connector.

1a. Solder the ESCs to the gearmotors

The Silver Spark motors have one tab near a red dot and another tab without one. This corresponds to the polarity of the motor. Each tinyESC **5** has one purple and one blue wire. Solder one of these wires (which

one doesn't matter at this point) to one tab then the other color to the other tab.

Do the same process to connect the other tinyESC to the other motor, but flip the polarity of the wires. For example, if you soldered blue to the tab near the red dot on one motor, then solder the purple wire to the same tab near the red dot on the other motor.

I recommend adding heat-shrink tubing to cover the joints from potentially shorting and to add strength to the connection.

1b. Solder the power connections

Solder all the red wires from the tinyESCs and the 9V regulator **6** to one of the tabs of the FingerTech mini power switch **1**. On the other tab, solder the red wire from the female JST connector. The female is the connector that the other connector goes inside of. This allows the (male) plug on the battery to plug directly into the connector, providing power.

The mini power switch works by tightening a screw which connects the two tabs. Once this screw makes contact the circuit is complete and electricity is able to flow. In order to make sure all components of the robot will turn off once the power switch is loosened, we connect all the red wires to one location, ensuring there is only one way for electricity to flow. We connect the other tab on the switch to the battery to ensure power is being received.

1c. Solder the ground connections

Solder all the black wires from the tinyESCs and 9V regulator to the black wire on the female JST connector. These connections don't need to be separated by the switch since we already separated the red wires. By connecting the black wires we complete the circuit, leaving only the switch to allow you to turn it on and off.

Congratulations. You've now soldered all of the most important connections in the robot!

2. CONNECT THE RECEIVER AND SERVO

Connect the receiver leads from the tinyESCs to channels 1 and 2 on the receiver **3**. These will allow you to control the drivetrain using the right

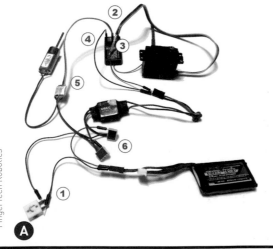

FingerTech Robotics

A

stick on your transmitter.

Connect the receiver leads from the servo ❷ to channel 3 on the receiver. This will allow you to control the lifting arm using the throttle channel on your transmitter.

Connect the receiver lead from the 9V regulator ❹ to channel 4 on the receiver. This allows the regulator to power the servo so it can use its full capacity.

3. BUILD THE WHEELS
3a. Glue hubs into the wheels
Add a dab of hot glue onto the larger half of a 3D-printed wheel hub and insert it into the center of a foam wheel (Figure ❶).

Flip the wheel over, add a dab of hot glue onto the smaller half of the hub (Figure ❸) and insert it into the center of the larger half. Press the pieces together to make sure the hubs properly bond to the foam wheel.

3b. Glue Lite Hubs into the hubs
Add a bit of hot glue to the nylon body of a Lite Hub then insert it into the center of the printed hub in the wheel (Figure ❹).

Repeat Step 3 for the second wheel.

> **TIP:** If there is difficulty inserting the Lite Hub, I recommend cleaning up the hub opening with a small drill bit or hobby knife.

4. MOUNT THE DRIVE MOTORS
Add a dab of glue to the bottom of each motor mount in the chassis (Figure ❺).

Now you'll heat the rear section of the chassis to insert the Silver Spark gearmotor. Turn on the heat gun and let it come up to temperature. Focus it on one motor well at a time: Once the plastic becomes soft, insert the drive motor into the slot. Ensure the gearbox is thoroughly mounted and sitting in the hot glue. Then bend the plastic in slightly so it helps support the motor. Remove the heat gun and let the plastic cool so it hardens in place (Figure ❻).

Repeat this process for the second gearmotor (Figure ❼).

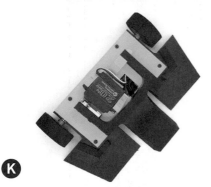

5. ASSEMBLE THE LIFTING ARM

Slide the lifting arm over its supporting arm on the top plate, as shown in Figure **H**.

Attach the *servo horn* to the output shaft of the servo. The servo horn is a component that connects to the output shaft's *splines* (like gear teeth), which allows you to transfer its power to something else. Slide the servo directly into its mounting area. It should sit snugly next to the lifting arm (Figure **I**).

Attach M3 bolts through the lifting arm into the servo horn (Figure **J**). Due to the servo horn's design you will need to insert one screw first then attach the second one after it. You may need to trim screws to fit exactly.

6. SCREW IT ALL DOWN

Screw the servo into its mounting holes in the top plate, using the four #6×½" Plastite screws.

Attach the wheels to the gearmotor shafts by tightening the Lite Hub setscrews using the 0.050" hex wrench.

Finally, attach the top plate to the robot chassis using the five #4×⅜" Plastite screws (Figure **K**). Enjoy your completed Kerfuffle!

READY TO RUMBLE!

So now that you have a completed machine, what do you do with it? First, practice with your robot. Turn it on, using the ³⁄₃₂" hex wrench to screw down the mini power switch. Use the transmitter's right stick to move the robot (front/back and left/right) and the left stick to move the lifting arm servo (up/down).

- **Practice driving.** Get the hang of how your robot drives, and become accustomed to how it moves — this will allow you to compete more effectively later on.
  - Do figure-8s — it's a great way to get used to the robot's steering as well as get a good read on its speed.
- **Practice fighting.** Because this machine has no dangerous spinning elements, you can easily build clones of it and hold mini sumo competitions where you have timed skirmishes and try to out-drive and out-fight each other.
  - To attack, approach a raised area on your opponent. Because Kerfuffle relies

REPLY MAIL

T NO. 187   LINCOLNSHIRE IL

AID BY ADDRESSEE

NO POSTAGE
NECESSARY
IF MAILED
IN THE
UNITED STATES

E IL 60069-9968

on lifting (Figure **L**), it needs to get underneath the other robot, so find areas on your opponent where the arm can get below to maximize lifting ability.

- Wait until you are firmly under the opponent to lift. If you try to lift too early it can leave you exposed to attack, so be patient before using the weapon.
- If Kerfuffle gets flipped over, use the lifter arm to flip it right-side-up again!

- **Fight at events!** Check robotcombatevents. com to find combat robot events in your area. Check when/where competitions are happening and see which ones are hosting "Plastic Ant" classes as part of their competition, such as MACRO's events held in Severn, Maryland (Figure **M**). I highly recommend going to events as this is where you will get connected to members of the combat robot community and further learn about the sport.

If you're looking for tips and tricks, or would just like to meet other combat robot enthusiasts, the Combat Robotics group on Facebook is where many of the builders (especially the ones from *BattleBots*) discuss robot designs.

With these tips you're well on your way to entering the world of combat robots! Happy building!

KEEPING UP WITH KERFUFFLE

Since I first shared this Version 2 design, Kerfuffle's shape has changed dramatically as competition pressed further development. You can see Version 3.0 (Figure **N**), Version 3.3 (Figure **O**), and newer designs by following my team Bone Dead Robotics on Facebook or @bonedeadrobotics on Instagram!

YOUR NEXT COMBAT ROBOT

Ready for kinetic weapons?

- First, build a protective test box or safety arena (see page 54).
- Try building Irkin (Figure **P**), my plastic antweight vertical spinner bot: grabcad.com/ library/irkin-1lb-plastic-ant-combat-robot-1.
- Or add new weapons to Kerfuffle, like my sawbot Version 3.1 (Figure **Q**)! *

Nathan Story

THE ROBOTEERS

The real magic of fighting robots? Discovering the community love

Kurtis Wanner

Bunny Sauriol

Jon Bennett

Brandon Young

Lucy Du

Emmanuel Carrillo

The thing about being a robot builder is, well ... we're odd. If you're thinking about building a robot, well ... you're odd too. But that's OK! The very best thing about robot events is that they're entirely populated by weirdos. Every race, religion, gender, political bent, whatever. But everyone in the building really loves making things.

In all my years, I've never encountered a hobby or a sport that has such wonderful people. Most of my current best friends are people that I've met through robot events. The toughest competitors turn into wonderful teachers and pals as soon as the match is over. Can you imagine a NASCAR team helping a competing team get their car ready for the event? Probably not. But in robot competitions, this sort of thing happens all the time! Contestants loan each other tools, help each other work, and do everything they can to help out the newcomer.

Between matches, you'll often see bitter rivals hanging out in the pits chatting or sharing a beer. Teams vying for the same medal will request to be set up next to each other because they're dear friends. Tools get swapped, opponents weld your broken bot for you, and lifelong friendships are made. You'll meet people from countries you've barely heard of and see robots that you've never dreamed of.

And you'll leave a richer person.

—*David Calkins*

LET'S MEET A CROSS SECTION OF THE ROBOT COMPETITION COMMUNITY:

KURTIS WANNER

- **LOCATION:** Saskatoon, Saskatchewan, Canada
- **ROLE:** I design and sell kits and components, organize events, and compete internationally
- **YEARS ACTIVE:** 19 years
- **LEAGUES:** Kilobots Combat Robot Events — kilobots.com/events

- **ROBOTS:** Dozens of insect-weight bots. My heavyweight: Crash 'n Burn, a multibot with flamethrowers
  SELF-BUILT? Yes for all. Designing is half the fun!
- **ORIGIN:** Saw *BattleBots* on TV and wanted to compete, so in 2003 I started a local event. There was nowhere to get parts so I started a store as well!
- **ADVICE:** Start with a Viper kit! It's the lowest barrier to entry and you'll learn everything you need to know to move on to custom builds and other weight classes. (Read more about the Viper kit in "Combat Kitbots," page 40.)

BUNNY SAURIOL

- **LOCATION:** San Jose, California
- **ROLE:** In the robot world I'm the person people come to when they need help getting connected to other roboteers. I go to at least one event a month, usually 2 or 3, and help everyone I can whenever or however I can.
- **YEARS ACTIVE:** 20 years
- **LEAGUES:** Robot Fighting League (RFL), Norwalk Havoc Robot League (NHRL)
- **ROBOTS:** Malice (250lbs), Grudge Frog (12lbs), Mouser (3lbs), Mini Malice (1lbs), Micro Malice (150g)
  SELF-BUILT? I built all my robots and I designed Mouser myself, but **Isaak Malers** designed Malice 250, Mini and Micro Malice, and Grudge Frog. We usually use the lower weight classes to prototype for the larger weight classes!
- **ORIGIN:** I got started in combat robotics as a way to bond with my mom, and I met my current fiancé through it as well.
- **ADVICE:** Start small, you can always work your way up!

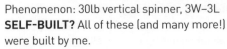

JON BENNETT

- **LOCATION:** Massachusetts
- **ROLE:** *BattleBots* photographer seasons 3, 4, and 6; competitor seasons 3 and 4; NHRL photographer
- **YEARS ACTIVE:** Since 2016
- **TEAMS AND LEAGUES:** Team Toad Combat Robotics, *BattleBots*, NHRL
- **ROBOTS:** Hypothermia and Texas Twister/ Spitfire (*BattleBots*)
  SELF-BUILT? I did assembly/welding on Texas Twister.
- **ORIGIN:** It's all Fuzzy's fault.
- **ADVICE:** Thompson's rule for first-time telescope makers also applies to robots: It's faster to build a 3lb robot, *then* a 30lb robot, than it is to just build a 30lb robot.

BRANDON YOUNG

- **LOCATION:** Bowie, Maryland
- **ROLE:** National competitor, weight classes from 150g to 60lbs
- **YEARS ACTIVE:** 11 years
- **TEAMS AND LEAGUES:** My own team is Bone Dead Robotics. In 2018, I founded the Leatherbacks Combat Robotics club at the University of Maryland, which invites students from different majors to learn about various skills relating to combat robotics such as design, manufacturing, and competing. I have been a competitor on the TV show *BattleBots* with the robots Big Dill and Mammoth.
- **ROBOTS:** Ferocious Mk.6: 1lb lifter, 6W–0L
  Kerfuffle: 1lb lifter, 25W–8L (build it on page 44)
  Demogorgon: 12lb undercutter, 11W–8L

Phenomenon: 30lb vertical spinner, 3W–3L
SELF-BUILT? All of these (and many more!) were built by me.
- **ORIGIN:** I got started in combat robotics after growing up watching the classic robot shows of the early 2000s (*Robot Wars*, *BattleBots*, and *Robotica*), which ignited a passion in me. As I grew up, I got more interested in the engineering behind the robots and continued to pursue that interest to this day.
- **ADVICE:** For people who want to get into the sport, I highly recommend attending local events (which you can find at robotcombatevents.com), seeing fights and talking to builders there. You can get a huge wealth of information and experience while minimizing cost. Smaller weight classes like 1lb and 3lb are best to start with; 250lb heavyweights are super expensive and are one of the hardest ways to learn lessons. 12lb and 30lb robots are also fairly pricey, so smaller machines are recommended.

LUCY DU

- **LOCATION:** Cambridge, Massachusetts
- **ROLE:** Design, machine, build, and compete!
- **YEARS ACTIVE:** 7 years
- **LEAGUES:** I compete at Norwalk Havoc Robot League as well as *BattleBots*.
- **ROBOTS:** HotLeafJuice: 12lb undercutter/ horizontal spinner, 18W–3L
  SawBlaze (2016–2021): 250lb overhead hammer-saw, 15W–8L
  Overhaul (2015): 250lb crushing lifter, 1W–2L
  SELF-BUILT? HotLeafJuice — designed/built with **David Jin**; SawBlaze — designed and built with a team.
- **ORIGIN:** Some of my friends were doing combat robots when I started graduate school, and it looked like a fun and interesting way to practice design/build/machining skills.

- **ADVICE:** Find a local event and compete! Smaller weight classes are easier to start with (and cost a lot less) and happen relatively frequently, so you can get a lot more practice iterating and driving your bot. Also, definitely talk to current builders (in-person or online) and ask questions; builders love talking about their bots.

EMMANUEL CARRILLO

- **LOCATION:** Seattle, Washington
- **ROLE:** I design robots ranging from 150g up to 250lbs. Mainly focused on making robots for *BattleBots* and making kits for new builders.
- **YEARS ACTIVE:** Started seriously building in 2015
- **LEAGUES:** Western Allied Robotics (a local Pacific Northwest robot group)
- **ROBOTS:** Big Dill: 250lbs lifter, 2W–3L WAR Hawk, 250lbs vertical spinner, 6W–5L MadCatter, 250lbs vertical spinner, 9W–6L **SELF-BUILT?** These were designed and built with a team. My smaller robots are all self-done, but the large robots require a team to really fund effectively.
- **ORIGIN:** Seemed like a fun sport that combined design, engineering, fabrication, and a bit of driving skill.
- **ADVICE:** Start small and quick. Be agile and adapt as you gain experience. You'll learn more at your first competition than you can ever read about. So just start. ●

Avery Wong Photography, Dan Longmire, Tony Woodward, Jamison Go, Emmanuel Carillo

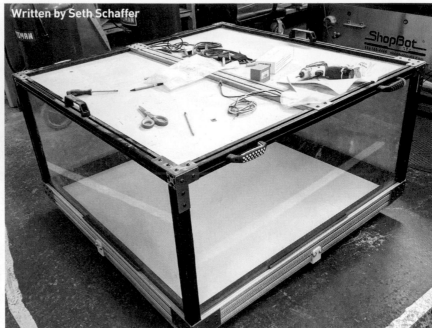

Written by Seth Schaffer

Build a DIY arena for safely watching robot mayhem

My ultimate goal with my business and my YouTube channel is to spread my love for the sport of combat robotics and to get others involved. So I designed an arena suitable for 1lb full combat antweights, plastic ants, and smaller bots with a completely free and open-source design for anyone to replicate, so more combat robot events can be held around the country and the world, safely and without too great an expense.

The entire arena plus streaming and safety gear can be built for under $1,000 without expensive or fancy tools. You can get the full Antweight Arena plans, bill of materials, 3D files, and construction guide for free or with an optional donation at justcuzrobotics.com/shop.

KEY FEATURES ARE:

- **6mm polycarbonate plastic (Lexan) walls/windows.** Never use acrylic (plexiglass), it is unsafe and will shatter.
- **80/20 aluminum extrusion** for the frame. This can be ordered cut to size by many suppliers.
- **Steel kick plates** to protect the Lexan from most common low-weapon hits.
- **Locking door** with top hinges and bottom latch.

- **LED light bars and cameras** for clear viewing, recording, and streaming. Position on the frame or ceiling as desired.
- **Leveling casters** with deployable feet provide mobility and a solid base for fights..

You can build this arena with little more than a cordless drill, some measurement tools, a hacksaw, and hex drivers. The toughest part is drilling holes in the steel kick plates, which would be greatly aided by the use of a drill press and quality drill bits.

Don't need a whole arena but still need to test dangerous kinetic weapons? Build a **test box** with a ¼" or 6mm polycarbonate top for as little as $70, following my video at youtu.be/qdSkZc4kfAw. ⊘

SETH SCHAFFER is a mechanical engineer at DEKA Research and Development, the founder of Just 'Cuz Robotics, builder of the feared beetleweight, Division, and a member of *BattleBots* Team Bots 'n' Stuff Robotics, builders of Bloodsport and the new 2021 robot, Retrograde.

Seth Schaffer

DAVID CALKINS is the founder of RoboGames. He also writes "Getting Involved" on page 22.

Written by David Calkins

HOW TO WIN AT ROBOT COMBAT
These 10 rules could turn you into a champ

Watching literally thousands of contestants cry at the smell of burnt metal and broken dreams, I created these rules for robot combat success. Follow them and you'll do well. If you don't, you won't.

1 KNOW CARLO BERTOCCHINI'S LAW AND LIVE BY IT

"Finish your robot before you come to the competition!"
If it isn't fully functional the week before the event, it's probably not going to pass safety or be able to fight. If your bot's not ready, volunteer on another team.

2 PRACTICE DRIVING

This is *the* most important thing I can say. The single greatest common denominator to winning is driving ability. Get that?

3 BE ABLE TO SELF-RIGHT

It is not a question of *if* your robot will be flipped over, it is only a question of *when* your robot will be flipped over. If you can't self-right, you'll never make it to the finals.

4 SIMULATE GETTING ATTACKED

If your robot cannot survive a good bashing with a sledgehammer, circular saw, and 10-foot free-fall, it will not last in the arena. Go nuts.

5 HAVE A WEAPON SYSTEM — OR MAYBE TWO

It's extremely rare for a wedge with no other weaponry to make it to the finals. Just make sure your weapon is allowable in the rules (no liquids, tasers, nets, or projectiles).

6 SIMULATE ATTACKING

Spend time in the garage or junkyard testing your weapon against a solid object. And make sure your bot is able to push double the maximum weight of your weight class.

7 GO TO A COMPETITION AND TAKE NOTES

You can learn more from other people's victories and mistakes than just your own.

8 USE GOOD BATTERIES, HAVE SPARES, AND MAKE SURE THEY'LL LAST 5 FULL MINUTES

If the batteries don't last the match, you're not gonna win.

9 DON'T LET THE JUDGES DECIDE THE MATCH FOR YOU

If the judges think the other robot was more aggressive and did more damage, then you're going to lose. There's only one way to absolutely ensure that you win: **Go for the knockout, every single match.**

10 READ THE DAMNED RULES

Read the rulebook from cover to cover. You need to do this for every competition — they are not the same between two events. ⊘

Read my unabridged set of rules, complete with gory details, at makezine.com/go/howtowinrobogames.

Hack Your Dreams

Written and photographed by Tomás Vega, Eyal Perry, Adam Haar, Oscar Roselló, and Abhi Jain

Build the Dormio "dream incubator" to influence and access your hypnagogic lucid dreams

TOMÁS VEGA, EYAL PERRY, ADAM HAAR, OSCAR ROSELLÓ, and **ABHI JAIN** are graduates of the MIT Media Lab in Cambridge, Massachusetts, with a mix of backgrounds in neuroscience, computer science, design, and biological engineering. For this project, Tomás led the hardware build, Eyal led software, Adam led science, Oscar led interaction design, and Abhi helped on all those areas — supported by the wonderful team at media.mit.edu/projects/sleep-creativity/people.

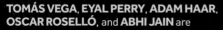

As you fall asleep, halfway between consciousness and unconsciousness, a window of opportunity opens for hacking your own dreams.

In this special period during the onset of sleep, called *hypnagogia*, simple stimuli like the movement of a car or the crackling of a campfire can change what you dream about. Our tool, the Dormio, helps you track this halfway period of semi-consciousness and then plays sounds to you at the specific moment when auditory stimuli are likely to redirect your dreams as you slip into sleep.

Historically, luminaries like Thomas Edison, Tesla, Poe, and Dalí each accessed this state of mind to capture creative ideas generated in their sleep onset dreams. Our own research suggests dreaming about a creative problem can improve your waking performance on it. And a recent neuroscience paper showed that people who access this hypnagogic sleep are able to triple their number of creative sparks when solving math puzzles.

It's exciting to think there are parts of our minds that we typically can't see, as they're hidden by sleep, and that we can use sleep tracking tools to uncover our own hidden creativity. If you decide to do this project it will have some tough spots, debugging, and improvising, but you will learn lots about sensors, sleep, and hopefully yourself.

WHAT: The Dormio V3 is the newest in a series of wearable devices we're making which help you track certain stages of your sleep and stimulate yourself with sound or smell to change your dreams in real time. We will make it together in this tutorial!

The Dormio V3 is a hand-worn device that tracks changes in three biosignals: heart rate, skin conductance (electrodermal activity or EDA), and finger muscle flexion. Changes in these signals indicate the *onset* of sleep, when there is a window of opportunity to play audio that will influence your dreams. Dormio broadcasts these three changing signals via Bluetooth to a web interface which detects sleep onset and plays audio during this special window of opportunity. It uses an nRF52832 Bluetooth microcontroller,

Oscar Roselló

TIME REQUIRED: A Weekend
DIFFICULTY: Moderate
COST: $125–$180

MATERIALS

» **Dormio V3 printed circuit board** $14 each from oshpark.com/shared_projects/YbsplJeh, or make your own PCB from the files at github.com/tomasero/openSleep, it's open hardware.
» **Electrical components, surface mount** The Fast Purchase list at Digi-Key (digikey.com/short/2ztm1m1m) includes: nRF52832 Bluetooth LE microcontroller, voltage regulator IC, battery charge controller IC, slide switch, 2.5mm stereo jacks (3), tactile switches (3), micro-USB connector, resistors, capacitors, inductor, crystal oscillator, and LEDs.
» **Heart rate sensor** $25 from pulsesensor.com
» **Flex sensor** Adafruit 1070
» **EEG electrodes** Amazon B06X1CL4S6
» **FTDI breakout board, 3.3V** SparkFun DEV-09873
» **JTAG/SWD debugger** Segger J-Link EDU Mini, Adafruit 3571
» **SWD cable breakout board** Adafruit 2743
» **Tag-Connect 6-pin "Plug-of-Nails" programming cable, "no legs" version** TC2030-IDC-NL, tag-connect.com
» **2.5mm stereo plugs with bare leads (2-pack)** Amazon B07ZT15JVM
» **Electrode lead wires with 3.5mm snap, 2.5mm plug** Amazon B07F5PXVWB
» **Electrical tape or heat-shrink tubing**
» **LiPo battery, 100mAh or bigger**

TOOLS

» **Soldering iron and solder, fine tip**
» **Magnifier**
» **Flux dispensing pen**
» **Wire strippers**
» **Computer with USB ports** to program your Dormio. Download the project code at github.com/tomasero/openSleep.

so we can use Adafruit's Bluefruit Feather BLE ecosystem to program it.

WHY: For decades it was accepted scientific fact that while we sleep our brains are turned off and we're not processing sight, sound, or smell. Instead, new evidence suggests our brain is not turned off in sleep at all — in fact many parts are more active than when we're awake, as they work to process memories from the day, restore cognitive function by clearing the brain's cellular waste products, and prepare for the next day by simulating possible future scenarios. We are conscious, dreaming, for most of the night, and our brain is still processing sound, sight, and smell enough that each of these sensory inputs can reliably alter people's dreams.

Even cooler, what we dream about changes how we think in the day. Our dream emotions carry over into our daytime emotions. Dreaming about something specific is tied to improved memory of that thing in the morning. Dreams can even augment creativity! That means dreams, with the right interface to influence them in targeted ways, can be a route to alter and improve your thinking.

WHO: You! Anyone and everyone can be a dream hacker. Even if you think you don't remember your dreams, this Dormio device will likely be a blast. People who typically forget their dreams can often remember them in hypnagogia.

Targeted *dream incubation* using Dormio is aimed at this early sleep stage, at night or in daytime naps, and wakes people directly during their dreams, so we especially encourage you to try this if you typically can't remember dreams or you think your dreams are boring. We bet you'll find some weird stuff in your head if you give it a go!

BUILD YOUR DORMIO DREAM INCUBATOR

1. ORGANIZE YOUR PARTS

First things first. Make sure you have your PCB, solder, soldering iron, flux pen, and electronic components ready to go. Your Dormio V3 board should look like Figure Ⓐ.

I like to put a piece of double-sided tape on the back of the PCB, just so it doesn't shift while being soldered. If you do, be sure to limit the time the soldering tip touches the board or the tape

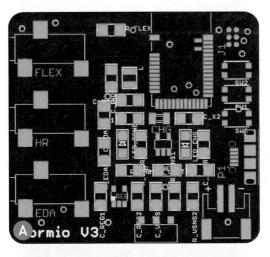

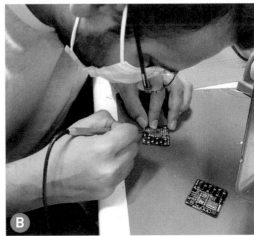

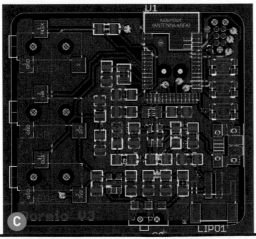

will melt. Solder on a silicone heat-proof mat if you have one (Figure **B**).

2. SOLDER COMPONENTS ONTO PCB

You'll need a PCB design software to click around and see which part goes where. First, download the board and schematic files from github.com/tomasero/openSleep/tree/master/hardware.

Then you can upload them to the free Altium Online Viewer (altium.com/viewer) to see how the components should be arranged. Or download Eagle, which offers a free version for hobbyists (autodesk.com/products/eagle/free-download). This is the software we used to designed the board and can also be used to visualize all the components and traces (Figure **C**). This way you can looking at the .brd and .sch files while you solder all the components in place.

Now you can start soldering the components! Use generous amount of flux when soldering the MCU and check closely to avoid bridging. In the timelapse video of Tomás working (youtu.be/FMHOy8AFdqA) he's using a stencil (and taking a long time to align it!) and solder paste, but this is not necessary. The board was redesigned so that you can assemble it using only a soldering iron, with no need for a reflow oven.

Once the board is all soldered up it should look like Figure **D**.

3. CONNECT THE PROGRAMMING GEAR

Check that you have the J-Link, FTDI, and SWD breakout boards.

Solder the included jumper pins onto the SWD breakout board.

Plug the J-Link into the SWD breakout board, then connect the J-Link to your computer by USB. Also connect the FTDI connector to your computer by USB (Figure **E**).

Connect the SWD and FTDI breakout boards to the Tag-Connect cable as shown in the diagram (Figure **F**). When powering the Dormio board using a battery, connecting the VCC of the FTDI breakout board isn't necessary.

Use the pinout diagram in Figure **G** as reference to match the pins of the Tag-Connect cable (Figure **H**). Note that the connections you attach should be mirrored from their layout on the PCB viewer (Figure **I**).

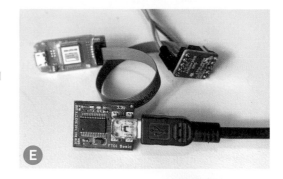

E

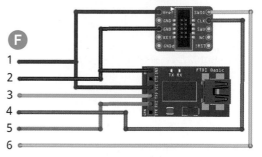

F

1
2
3
4
5
6

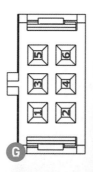

G

H

RX(5)
TX(3)
SWDIO(6)
SWCLK(4)
GND(2)
Vcc(1)

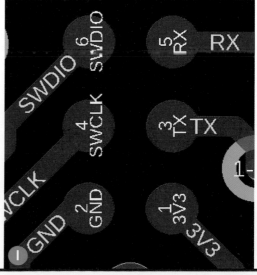

I

J

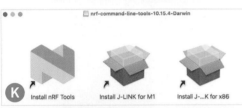

K Install nRF Tools   Install J-LINK for M1   Install J-...K for x86

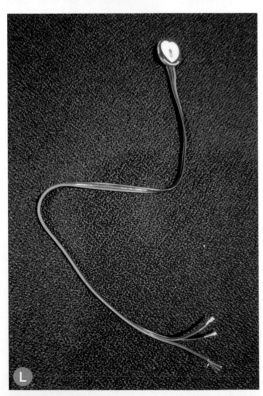

L

You can now connect the Tag-Connect to the Dormio V3 PCB (Figure **J**).

4. PROGRAM YOUR DORMIO
Now on your laptop, download and install the nRF command line tools from the Nordic website (nordicsemi.com/Products/Development-tools/nrf-command-line-tools/download). From this nRF package, also install the J-Link drivers for your machine (Figure **K**). Open a terminal window and type **nrfjprog -v** to verify that the nRF tools and J-Link driver are installed.

Download the Adafruit board files to Arduino IDE, following the steps at learn.adafruit.com/bluefruit-nrf52-feather-learning-guide/arduino-bsp-setup. Also grab the Adafruit bootloader from github.com/adafruit/Adafruit_nRF52_Arduino/tree/master/bootloader/feather_nrf52832. Then run these commands in terminal to program the bootloader:

```
nrfjprog -f nrf52 --eraseall
nrfjprog -f nrf52 --program
feather_nrf52832_bootloader-0.6.2_
s132_6.1.1.hex
nrfjprog -f nrf52 --run
```

Alternatively you can flash the bootloader by forcing the device in bootloader mode, by following the tutorial at learn.adafruit.com/bluefruit-nrf52-feather-learning-guide/using-the-bootloader.

Finally, grab the Dormio firmware file *sleepduino_nrf52.ino* from Tomás' GitHub, github.com/tomasero/openSleep/tree/master/embedded/sleepduino_nrf52. In the Arduino IDE, select FTDI on the Serial Port, then hit Upload to flash the firmware to the board. Now your Dormio is programmed!

5. MAKE THE SENSOR AUX CONNECTORS
The Dormio V3 has three 2.5mm stereo aux connector jacks, just like you use with headphones. We use a 2.5mm male plug for sensor connection, so you can take the sensors in and out with ease.

You just need to connect each sensor to a male aux cable and you're good to go! The aux cables have a signal (white), ground (black) and power (red) wire each exposed.

The heart rate sensor will use all three wires.

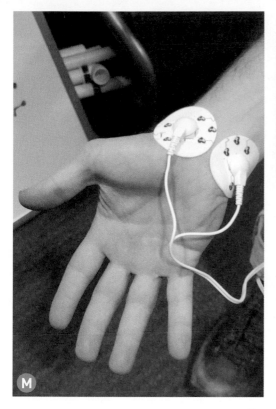

Solder each wire of your Pulse Sensor (Figure ⓛ) to the respective wire of an aux cable: red to red (power), black to black (ground), and purple to white (signal).

For EDA sensing, the electrode cables already have their own 2.5mm plugs (one wire connects to power and one to signal). To connect the electrodes, you only need to snap them onto the snap cable heads, and then you can stick the electrodes on your wrist when it's time to sleep (Figure ⓜ).

For the flex sensor, solder one lead to an aux plug signal wire (white) and the other to ground (black). Once you have the wires soldered to the flex sensor, we recommend you wrap some electrical tape around the base of the sensor to give it some rigidity at the base (Figure ⓝ). Another alternative is to use heat-shrink tubing.

6. MAKE IT WEARABLE
You can really attach this device to your hand however you want to. The only necessities are that the heart sensor stays on your fingertip, the electrodes stay on the inside of your wrist, and

the flex sensor stays attached to your finger.

An earlier version of the Dormio used a glove (Figure ⓞ), but then people complained their hand was too hot while they slept.

A newer version uses simple velcro straps for

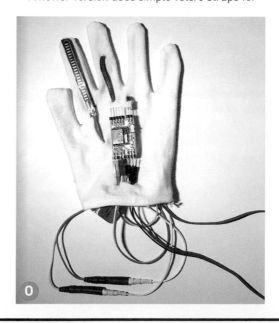

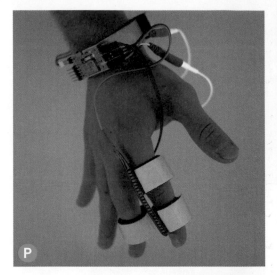

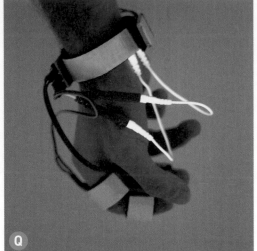

Oscar Roselló

a more breathable attachment setup (Figures P and Q).

Here is a version made by Gershon Dublon, Xin Liu, and Clement Duhart (Figure R) based on our open source plans. They 3D printed a circuit case!

The easiest solution is just to attach a bit of velcro to the back of the PCB you've assembled (hot glue works well), wrap another piece of velcro around your wrist, and attach the PCB to that. You can use velcro or adhesive to attach the battery under the board. Now you just secure the sensors to your fingers (velcro again works well) and you're good to go!

7. CONNECT TO THE SITE

Now all you have to do is connect using Chrome web browser (Chrome >v56 for Mac, Chrome >v70 for Windows), either by locally opening the file serverless/wearable.html from the GitHub repo, or by going to our online web interface at eyalperry88.github.io/dormio/wearable.html.

Power up your Dormio by plugging in the battery. Since the board is programmed, you'll

start seeing signals on the web page right when you plug in the sensors. Check out Tomás testing his board and sensors at youtu.be/osPBl1xgNzU.

The web interface (Figure S) is designed to:
- Connect to your Dormio sensor device
- Record your dream incubation audio stimulus, and also an audio message to wake you up and ask for a dream report
- Sense real-time changes in your biosignals
- Play the dream incubation audio when you slip into the beginning of hypnagogia
- Play the dream report audio when you near the end of hypnagogia
- Repeat this cycle as many times as you want, for multiple rounds of hypnagogic dreams!

Here's a clickthrough demo of how to use it: youtu.be/oxPSyNIO6tw. Record your dream incubation stimulus, and lie down for a nap while wearing your DIY dream incubation device!

You'll notice that at the end of your session, your data and audio are downloaded to a local *.zip* file, not uploaded to any server where anyone else can see it. Your dream, your incubation, your

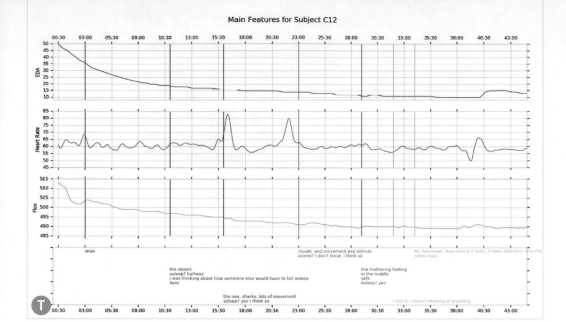

Main Features for Subject C12

audio — private.

If you want to, you can plot your biosignals to see changes as you slip into sleep. Figure **T** shows what a typical descent into hypnagogia with wakeups and dream reports looks like.

DREAM ON!

You now have a basic interface that will let you change your dream content! We suggest you use this device during a nap, and that you make sure to set a clear intention to dream about your chosen dream theme before you lie down, even saying out loud why you want to dream of that theme.

You may have to tweak things to get this whole hypnagogic incubation practice right — from your bedtime to the volume of your audio — because this is not an exact science. But though it is inexact, it is an ancient, important, human practice; dream incubation has been used for at least 4,000 years to seek answers to life's questions, to find creative inspiration, and to heal.

These are your dreams, and this is your dreamscape, so explore what you will! Maybe you want to have dreams of Batman, and Beyoncé, and more power to you. Maybe you want to travel inside your favorite video game, or to your favorite planet, all in your own imagination. Please, do tell us how it goes.

GOING FURTHER

Dormio V3 is only built for tracking and changing sleep onset, a tiny slice of the overall experience of sleep and dreams. Want to go further?

- Try and build something that tracks a different sleep stage (maybe an eye tracker that can capture REM sleep) and see if you can change your dreams in a later stage. We built a different device aimed at REM that looks like a sleep mask (media.mit.edu/projects/masca), if you want inspiration, and wrote a paper about it for a class project (ethanweber.me/sleep/Human_2_0_Paper.pdf).

- Think more about the comfort and attachment of the Dormio device on the hand. Build a better housing or a better PCB fixture than velcro. We'd love to see your ideas and inspirations, let us know at dormio@media.mit.edu! ⊘

MORE ON DORMIO AND LUCID DREAMING

- The Dormio team, ongoing work, and published papers: media.mit.edu/projects/sleep-creativity
- How to build Dormio V2 (previous version, with lots of overlap). Click on "Input Device": fab.cba.mit.edu/classes/863.17/CBA/people/tomasero/index.html
- How the sleep data streaming interface was made: same page, click on "Interface and Application"
- *Demystifying Medicine* video on how Dormio works and what it's used for: youtu.be/9dGuxNb9mCY
- News report on Dormio's potential for dream incubation: news.mit.edu/2020/targeted-dream-incubation-dormio-mit-media-lab-0721
- Ethics of incubating dreams with devices: aeon.co/essays/dreams-are-a-precious-resource-dont-let-advertisers-hack-them
- Lucid dreaming and dream incubation: dreamscience.ca/en/dreamfaq.html

ZenBot

Build a self-playing meditation drum actuated by seven 2-axis mallet-wielding robots

Written by David Covarrubias

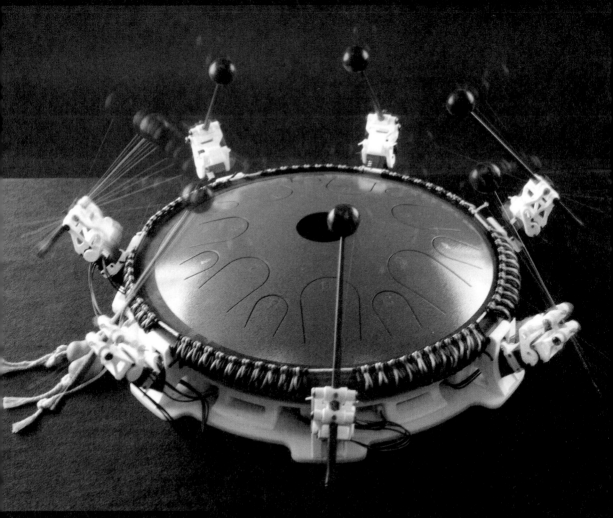

 DAVID COVARRUBIAS is an animatronic effects designer whose 28 years of film credits include *The Lost World*, *Iron Man*, *The Mandalorian*, *Suicide Squad*, and many more.

BUSINESS REPLY MAIL

FIRST-CLASS MAIL   PERMIT NO. 187   LINCOLNSHIRE IL

POSTAGE WILL BE PAID BY ADDRESSEE

Make:

PO BOX 566
LINCOLNSHIRE IL 60069-9968

When things break around the house it usually means a trip to the hardware store for replacement parts, but when my doorbell chime stopped working, I saw it as an opportunity to use some animatronics skills and make something more fun and aesthetically pleasing than a standard "ding-dong" type doorbell. This is how ZenBot was born.

ZenBot is an Arduino device that, by means of seven mallet-wielding 2-axis mini robots, can play soothing percussive tones on a meditation drum, aka tongue drum.

While ZenBot was meant to be a basic doorbell, it has potential to be expanded into a MIDI instrument by using available open-source libraries, or even downsized to work with a smaller drum in order to significantly reduce material costs.

In this article I'll cover my design process while also providing enough explanation and online resources for you to build one of your own. Along the way you'll learn about design organization, setup and use of Dynamixel robotic servos, and beginner-level Arduino programming. Before we jump in, check out the video of ZenBot playing Tears for Fears' "Mad World" at makezine.com/go/zenbot. You're going to want to build this.

DESIGN GUIDELINES

As a starting point, I set a few guidelines and limitations for myself in order to narrow down the options:

- Utilize Robotis XL330-M288-T servos for their low cost and programmability — and because I already owned a box of 25 of them that were longing to be used.
- Utilize a specific 14" tongue drum which I had also recently purchased.
- All parts must be printable on my Ultimaker 2 Extended 3D printer. (All but one ended up being done this way.)
- Design must be modern looking yet work well with the more traditional feel of the drum itself.
- Design must be axially symmetrical in nature.
- Design must mimic human hand motion using rubber mallets. For a while I considered using solenoids, but swinging mallets just felt like a more fun and organic approach.

TIME REQUIRED: 1–2 Weeks
DIFFICULTY: Intermediate
COST: $350+

MATERIALS

- » **Tongue drum, 14"** aka hand pan drum or meditation drum. I used eBay #233595745739.
- » **Drum mallets, 1" ball end (7)** Get two 4-packs of wooden mallets, eBay #184510420465, or 4 pair of rubber mallets, Timber Drum Co. #TMD2.

FROM ROBOTIS (robotis.us**):**
- » **Servomotors, Dynamixel XL330-M288-T (14)**
- » **Robot cables, X3P type:** 240mm (10), 100mm (20), and 180mm convertible (10)
- » **U2D2 servo programmer**
- » **Power Hub Board adapter for U2D2**
- » **Microcontroller, OpenCM9.04-C**
- » **Expansion board, OpenCM 485**

FROM AMAZON:
- » **Power adapter, AC to DC** AC input 100V–240V 50/60Hz 1.0A, DC output 6V 3A, #B07N18XN84
- » **Male headers, 0.1", 40-pin breakaway (1pk)** #B015KA0RRU
- » **Potentiometers, 10kΩ, linear taper (5)** #B09MS1T32Z

FROM MCMASTER-CARR:
- » **O-ring cord, Buna-N rubber, 0.07" wide, white (10')** #3286T11
- » **Machine screws, button head 6-32, ½" long (28)** #97763A143
- » **Machine screws, flat head 6-32, ⅜" long (21)** #93791A453
- » **Machine screws, button head 10-32, ½" long (28)** #97763A232
- » **Machine screws, socket head 4-40, ¼" long (14)** #92610A108
- » **Cast acrylic sheet, 24"×24"×¼", white** #8505K757

TOOLS

- » **Computer with USB-micro cable, Arduino IDE, and Dynamixel Wizard software** free online from arduino.cc/en/software and emanual.robotis.com/docs/en/software
- » **3D printer** with Tough PLA or similar filament
- » **Hot glue gun**
- » **Drills, taps, reamers, and countersink** for finishing holes
- » **Basic electronic tools: soldering iron, multimeter, wire, heat-shrink tubing, etc.**
- » **Sandpaper, files, hobby knives, etc.** for cleaning 3D-printed parts
- » **Laser cutter (optional)** or laser cutting service, to cut the base out of acrylic. Base can also be made from wood, or 3D printed in smaller sections.

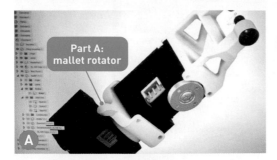

Part A:
mallet rotator

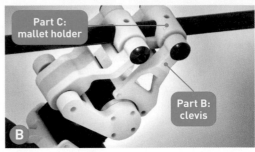

Part C:
mallet holder

Part B:
clevis

BUILDING THE ZENBOT

After coming up with a plan, I broke the project down into several parts. All seemingly complicated projects are just a combination of many much simpler components. In the case of the ZenBot, this is what they ended up being:

1. ROTATION MECHANISM

With 14 notes on the tongue drum, I decided to build 7 identical robots, each able to hit a note on the larger tongue and then pivot to the smaller tongue to hit the same note an octave higher.

Dynamixels are great for their modularity. Designing a rotation move was as simple as creating a bracket, Part A, that could be bolted onto the servo horn of the base servo and also onto the body of the second servo (Figure **A**). On a design with stronger forces acting on the joint, this approach would've been inadequate, but in this scenario, the simple approach was sufficient.

2. STRIKE MECHANISM

Designing the strike mechanism was a little more involved. It required deciding what paths I wanted the mallet to travel within. After determining this, I was able to pick the axis of rotation for the pivot, as well as the second axis that the stick would rotate around as it swung up and down. Once I had the pivot positions figured out, I designed a lightweight structure that supported the clevis joint, Part B, as well as the mallet holder, Part C (Figure **B**).

3. ANTI-DAMPENING MECHANISM

I was afraid that if the servo couldn't change directions fast enough after hitting the drum, it would result in a muted and not very pleasant-sounding tone. The mallet needed to strike the drum hard but also retract fast enough to allow the tongue to resonate. I solved this in the simplest way that popped into my head: Part D. The mallet would be pushed by a stretched rubber band. The band would stretch a bit on the way down, then release the force onto the mallet like a slingshot. The servo would already be retracting as the mallet hit, then the rubber band

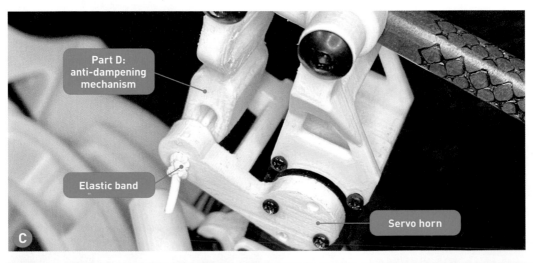

Part D:
anti-dampening
mechanism

Elastic band

Servo horn

would catch the mallet on the way up, thereby keeping it from dampening the sound or double bouncing on the drum (Figure C).

But this method would require finding that perfect synchronization between the elastic, the mallet, and the servo position in order to work. I wrote an Arduino sketch that allowed me to use potentiometers to dial in the position when the mallet is down, position when the mallet is up, and the amount of time in `millis()` (milliseconds) between when the down command and the up command are sent. Once I had the numbers needed to get the loudest and cleanest sounding note, I just plugged them into a function that handles the mallet strike of each robot.

4. SUPPORT STRUCTURE

After part cleaning, thread tapping, and assembly, I had a complete mallet assembly which, after a successful test, was duplicated 6 more times to cover the rest of the notes on the drum.

The next step was a solid surface to mount them onto. The base, Part E, was designed to not only hold the servo, but to also help capture the edge of the drum to keep it held down in the center of the circle. Care was given to allow clearance for screwdrivers and screws to reach their mark as well as the servo cables that needed an exit path.

At this point, I still had 7 individual robots with nothing keeping them grounded at the proper location relative to the drum. I designed some bridge parts, Part F, to help fix the robots in their

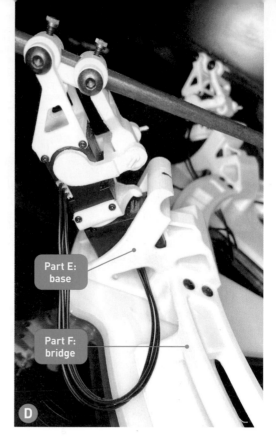

Part E: base

Part F: bridge

D

proper place and orientation (Figure D).

Now the servos were bolted together into a single assembly, but it needed a little more support, so a piece of acrylic was laser cut to become the foundation, Part G, that everything was then bolted onto (Figure E). I could have designed it as 7 pieces that could be grown on a 3D printer, but because I had a laser cutter available, the simplest solution was to laser cut a single part and bolt everything onto it.

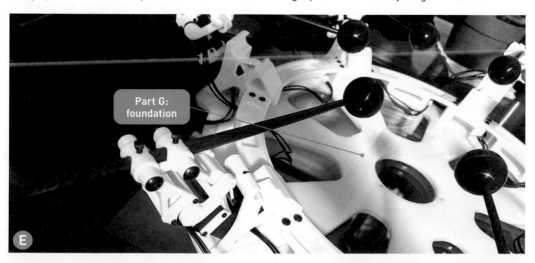

Part G: foundation

E

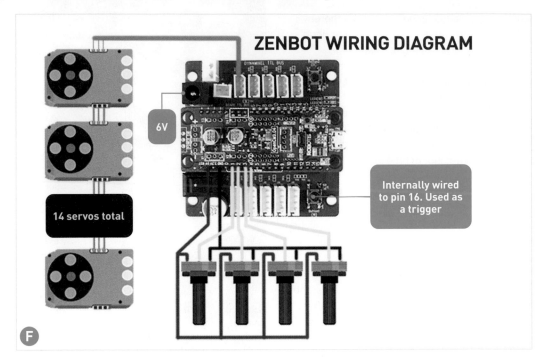

ZENBOT WIRING DIAGRAM

6V

14 servos total

Internally wired to pin 16. Used as a trigger

F

5. WIRING THE CIRCUIT

Wiring these Dynamixel servos is very easy since they can be daisy-chained together. It's important to note that they are all factory assigned ID number 1. So I used the U2D2 programmer and Dynamixel Wizard app to assign each servo a unique ID number. I chose to label the rotation servos with ID's 1–7 and the corresponding mallet strike servos as 11–17.

At this point I simply created a chain going from 1 to 11, 11 to 2, 2 to 12, and so on. The last servo got a longer extension and plugged into the Arduino-based OpenCM 9.04 microcontroller with expansion board (Figure **F**). The expansion board allows for other types of Robotis servos to be used and has a nice power switch and DC input plug.

For this device I simply added a pushbutton switch to represent the doorbell trigger. It is possible to incorporate a wireless switch to trigger the sequence, but for me a wired switch was able to do the job just fine.

6. PROGRAMMING

I figured out early on that the drum when hit by the mallet will just ring on, so the start of the note matters here, but its end is not controlled.

This means that a 16th note and a quarter note will sound the same. This eliminated the need of complicated code to handle timing.

The general idea was to create seven arrays, one for each robot. Within the array, each piece of data represented one step in time, and the value determined what the robot would do. A value of **0** doesn't play any notes, while a **1** will play the low note, and a **2** will play the octave note on the smaller tongue (Figure **G**).

If the start button was pressed, the robot will play a low note, high note, or rest, as per the first data value in each array. It will then look ahead and reposition itself based on the next note that needs to be played (Figure **H**). When it is time, the next note will play, and the process will continue until all the notes in the array are played. The tempo is governed by a simple delay executed after each note packet is executed.

This approach works great if you are creating a "set it and forget it" type of device, such as a doorbell. Filling the arrays with notes can be a little time consuming, but once it's populated, you're done. For any other application, perhaps turning it into a MIDI instrument might be more beneficial.

```
ZENBOT_MadWorld_MAKEMAG

// 0= no note  1=low note loud  2= high note loud
int notesC[160] = {0, 0, 0, 0, 0, 0, 0, 0, 0, 0, 1, 0, 0, 0, 1, 0, 1, 0, 0, 0, 1, 0, 0, 0, 0, 0, 0, 0, 0, 0, 0,    0, 0, 0, 0, 0, 0,
int notesD[160] = {1, 0, 1, 1, 0, 0, 2, 2, 0, 0, 0, 0, 0, 0, 0, 0, 0, 0, 0, 0, 0, 0, 0, 0, 0, 0, 1, 0, 0, 0, 1, 1,    0, 0, 1, 1, 0,
int notesE[160] = {0, 0, 0, 0, 0, 0, 0, 0, 0, 0, 0, 0, 0, 0, 0, 0, 0, 0, 0, 1, 0, 0, 1, 1, 0, 0, 0, 0, 0, 0, 0, 0,    0, 0, 0, 0, 0,
int notesF[160] = {0, 0, 1, 0, 2, 2, 1, 0, 1, 0, 0, 0, 1, 2, 0, 0, 0, 0, 0, 0, 0, 0, 0, 0, 0, 0, 0, 2, 0, 0,    0, 0, 1, 0, 2, 2,
int notesG[160] = {0, 0, 0, 0, 0, 0, 0, 0, 0, 0, 0, 0, 0, 0, 0, 2, 2, 1, 2, 0, 0, 1, 0, 2, 2, 0, 2, 1, 0, 0, 0,    0, 0, 0, 0, 0, 0,
int notesA[160] = {0, 0, 1, 0, 0, 0, 1, 0, 2, 2, 1, 2, 0, 0, 1, 0, 0, 0, 0, 0, 0, 0, 0, 0, 0, 0, 0, 0, 0, 0, 0, 0,    0, 0, 1, 0, 0, 0,
int notesB[160] = {0, 0, 0, 0, 0, 0, 0, 0, 0, 0, 0, 0, 0, 0, 0, 0, 0, 0, 0, 0, 0, 0, 0, 0, 0, 0, 1, 0, 0, 0, 1, 0,    0, 0, 0, 0, 0, 0,
```

```
for (int i = startFrame; i < endFrame + 1; i++) {
  // STRIKE NOTE
  if (notesC[i] == 1 || notesC[i] == 2)  dxl.setGoalPosition(11, malletDown);
  if (notesD[i] == 1 || notesD[i] == 2)  dxl.setGoalPosition(12, malletDown);
  if (notesE[i] == 1 || notesE[i] == 2)  dxl.setGoalPosition(13, malletDown);
  if (notesF[i] == 1 || notesF[i] == 2)  dxl.setGoalPosition(14, malletDown);
  if (notesG[i] == 1 || notesG[i] == 2)  dxl.setGoalPosition(15, malletDown);
  if (notesA[i] == 1 || notesA[i] == 2)  dxl.setGoalPosition(16, malletDown);
  if (notesB[i] == 1 || notesB[i] == 2)  dxl.setGoalPosition(17, malletDown);
  delay(time2Move);

  //PREP FOR NEXT NOTE
  if (i + 1 >= endFrame + 1) a = startFrame; else a = i + 1;
  if (notesC[a] == 1)dxl.setGoalPosition(1, malletRotateLowNote[1]); else dxl.setGoalPosition(1, malletRotateHighNote[1]);
  if (notesD[a] == 1)dxl.setGoalPosition(2, malletRotateLowNote[2]); else dxl.setGoalPosition(2, malletRotateHighNote[2]);
  if (notesE[a] == 1)dxl.setGoalPosition(3, malletRotateLowNote[3]); else dxl.setGoalPosition(3, malletRotateHighNote[3]);
  if (notesF[a] == 1)dxl.setGoalPosition(4, malletRotateLowNote[4]); else dxl.setGoalPosition(4, malletRotateHighNote[4]);
  if (notesG[a] == 1)dxl.setGoalPosition(5, malletRotateLowNote[5]); else dxl.setGoalPosition(5, malletRotateHighNote[5]);
```

BUILDING YOUR OWN

To build your own ZenBot, go to the project page at makezine.com/go/zenbot to download the Fusion 360 files for printing and the Arduino code for the microcontroller, and to watch my video tutorials that will take you through the build in fine detail.

Take care to measure your components well. Drums and mallets that are handmade are not always built to exact measurements. If yours are a bit different than mine, you might need to make minor adjustments to the robot parts. This is where the Fusion 360 files will come in handy.

TAKING IT FURTHER

Now that you have the skills to build your very own ZenBot, here are some ideas for taking it to the next level:

• Convert the code to accept live input from a MIDI keyboard, or maybe turn it into a modern MIDI-controlled player piano-type instrument.
• Experiment with the anti-dampening mechanism or create your own version. Maybe try a flexible/compliant mechanism instead of the elastic band to make it more durable and use fewer purchased parts.
• Make the trigger button wireless. Maybe use Bluetooth or RF to trigger the drum from a distance.
• Design a smaller, low-cost version by using

a smaller drum with fewer tongues. Reduce the servo count from 14 to 7 by eliminating the need to rotate.
• Or take what you learned here and apply it to a xylophone, tubular bells, chimes, or any other melodic or percussive instrument.

No matter what you do with it, one thing is for sure, you will have a beautiful functional art piece that is sure to be a great conversation starter! ✪

Matt Alavi

Inductive Adornments

Send electrical power through the air to light up LEDs wirelessly

Written by Lee Wilkins

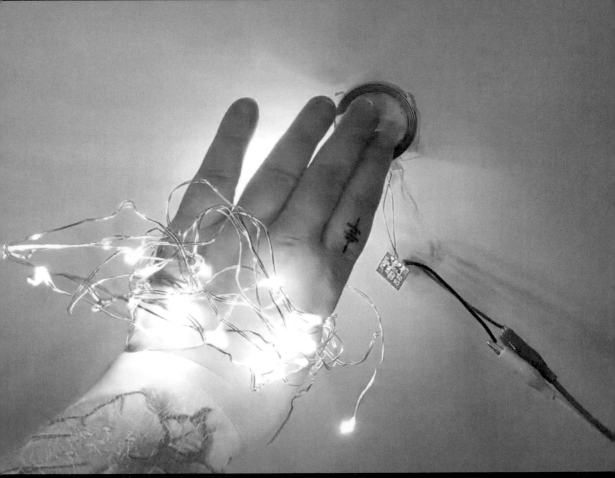

LEE WILKINS is an artist, cyborg, technologist, and author of our new "Squishy Tech" column in *Make:* looking at technology and the body and how they intertwine. Follow them on Twitter @leeborg_

Inductive chargers are becoming more popular for phones, watches, headphones, and other devices — they're everywhere! I look forward to imagining a completely wireless future, where power moves through the air like Wi-Fi and radio. But for now, I'll show you how to use induction to make some fashionable wireless accessories!

HOW DOES IT WORK?

At its core, wireless charging uses the magnetic field created by every electrical wire to transfer power between coils. This is possible because of *Faraday's law of electromagnetic induction*, which tells us that an electrical force is produced across an electrical conductor (a wire) in a changing magnetic field. Faraday's theory is also responsible for important things like transformers, electric motors, and generators.

Every wire with an electrical force produces a magnetic field with a north and south pole, which can be further described with *Lenz's law*, which tells us how we can find the magnetic polarity of every electrical field. As shown in Figure **A**, you can predict the flow of the magnetic field by pointing your right thumb in the direction of the current; your fingers will then follow the flow of the magnetic field! It sounds wild, but I literally mean that every wire with electricity in it produces a magnetic field, which can be used to make things like motors, solenoids, and loads of other tech.

Because every individual wire's field is very small, adding many wires in a coil multiplies the strength of the force. The shape of the coil, as we'll explore, affects the shape of the field. Figure **B** shows a spring coil, but they can be flat, circular, helical, and many other shapes.

There are two coils called *inductors* in our wireless power circuit, one in the *transmitter* (the device providing power) and the other in the *receiver* (the device being powered). When a current passes through a coil in the transmitter — the charging station, for example — the moving electric charge creates a magnetic field, which passes through the opposite coil in the receiver, inducing a current in the receiver coil which in turn passes through a rectifier and goes on to charge your battery or power your device.

Inductors are typically made of enamel-coated

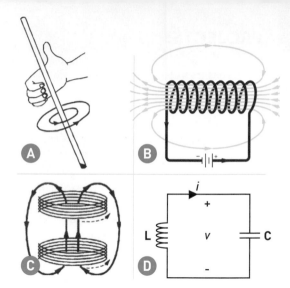

copper wire, aka magnet wire, because copper has low resistance and the enamel insulation ensures the current passes through the entire spiral in the coil, and doesn't short-circuit across it.

COUPLING INDUCTORS

There are a few factors that govern how far you can hold your transmitter coil from your receiver coil, as well as the power capacity for your coil. Inductors transfer power by *coupling*, meaning the two electromagnetic fields mesh together. Playing with the parameters below can help you achieve this more efficiently. Electromagnetics and inductive circuitry are a whole field of study, so I'll just give you the basics here that can help you get started.

- **Inductor size and shape** have a direct impact on coupling. For the best results, as many lines of the magnetic field produced by the transmitter coil must intersect directly with the receiver coil. Pairing coils of the same shape will ensure the best coupling (Figure **C**).
- **Distance** is also an important factor in power transfer. As the coils move apart, the inductive coupling is rapidly reduced; efficiency of 90% or more can only be achieved if the distance-to-coil-diameter ratio is less than about 0.1, otherwise the efficiency falls rapidly. However, this can produce some interesting effects if you want your lights to fade!
- **LC circuits**, named for their components — an inductor marked L, and a capacitor marked C — can be used to create a more efficient coupling circuit (Figure **D**). Tuning the frequency of an LC circuit can create

a *resonance* frequency between the two inductors, increasing efficiency.

I encourage an exploratory approach to learning about inductive power. Most components are cheap and easy to set up. While the math is pretty complex, the ideas can be observed clearly with some simple LED circuits that I'll show you now.

YOUR FIRST INDUCTIVE BLINK

The first thing I like to share with people is the easiest way to make your own inductive blink, using just a piece of wire, an LED, and your phone! Because your phone has an inductor inside for NFC (Near Field Communication) for Apple Pay, Google Pay, or similar services, you can create a very basic inductive circuit.

MATERIALS
» **Any 5mm or 3mm LED**
» **Enameled copper wire** aka magnet wire; or any wire that's coated

TOOLS
» **Soldering iron**
» **Modern smartphone**

1. CHECK COMPATIBILITY
Make sure you have a smartphone that not only has NFC, but also can run NFC in the background. Starting with the iPhone 7, iOS 14, you will have an Enable NFC option. For most Android phones you can enable NFC by going into Settings→

Connected Devices→NFC and making sure it's switched on.

2. MAKE A COIL
Make a simple coil by wrapping your wire around a large marker. I'd recommend between 10–20 turns, depending on the gauge of your wire. Make sure to leave both ends of the wire accessible (Figure **E**).

3. SOLDER YOUR LED
Solder each lead of your LED to one end of your coil (Figure **F**). Polarity is not important here, so don't worry about the positive and negative leads.

4. LIGHT IT UP!
With the NFC setting on, place your phone over your LED coil (Figure **G**). You should see the LED start to blink! You might have to move your coil around, depending on your phone. You can also try flipping the coil backward to get it to couple.

EXPLORING INDUCTIVE POWER
If you're looking to get up and going with inductive LEDs quickly, I'd recommend getting Adafruit's **Small Inductive Coil and 10 Wireless LED Kit** (Figure **H**). You'll also need a 5V power supply: either a USB battery, a battery pack of AA batteries, or a bench power supply. (The bench supply is great for learning, but you'll need to figure out another option if you want to take your creations wireless.) Adafruit also sells a **Large Inductive Coil** wireless LED set, and a paired **Inductive Charging Set** in 3.3V

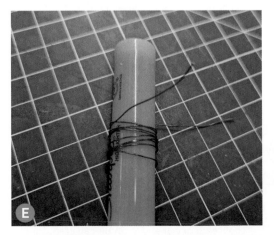

E

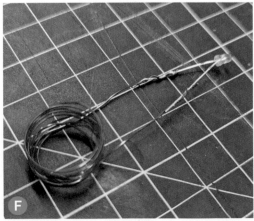

F

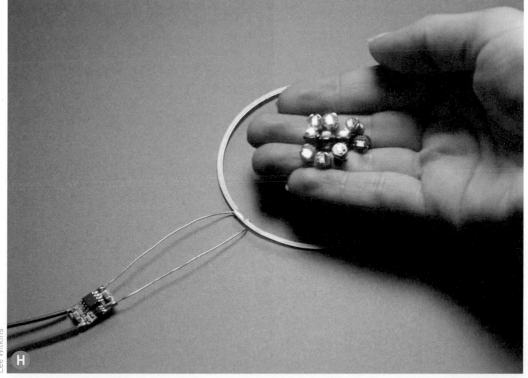

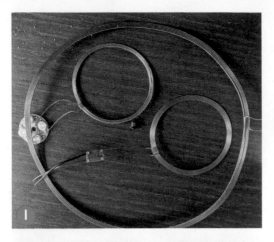

or 5V (Figure I), which you can use to power anything by connecting the receiver output coil to whatever you want. Once you're comfortable with these out-of-the-box solutions, you can start experimenting a bit more.

My go-to for this type of exploration is always AliExpress, where you can find a wide variety of transmitters and receivers at different voltages and sizes (Figure J). I recommend experimenting with both transmitter and receiver coil shapes; you'll find they all react differently.

- **Receivers** — As we've seen, you can make your own coils but premade coils have much more even and consistent winding. I've tried several winding jigs and can safely say there's nothing like a professionally machine-wound coil. These coils come in SMD, through-hole, and individually wound formats with a variety of sizes and shapes. They're pretty affordable, so I recommend buying a few and playing.

  Anything you want to power, you can solder

to your receiver coil: an LED, a motor, even a small circuit board. Rather than using large 5mm or even 3mm LEDs, I use 1206-size surface mount LEDs that provide a much smaller point of light. One of my favorite techniques for DIY coils is to embroider the coil onto a textile, like these from Kobakant (Figure K). Simply solder an LED to the ends of the coil as you did in the previous example.

- **Transmitters** — These are a bit more limited. I've had some luck with experimenting by cutting off and reattaching larger coils. It's important to keep an eye on the voltage rating for your transmitter. Larger coils might require more power to get the results you're looking for. I've had best results with these AN-100 radio loops made for AM radios (Figure L), soldered to a 12V, long distance transmitter. I think they're great for inductive adornments because these coils are beautiful and can fit over your head!

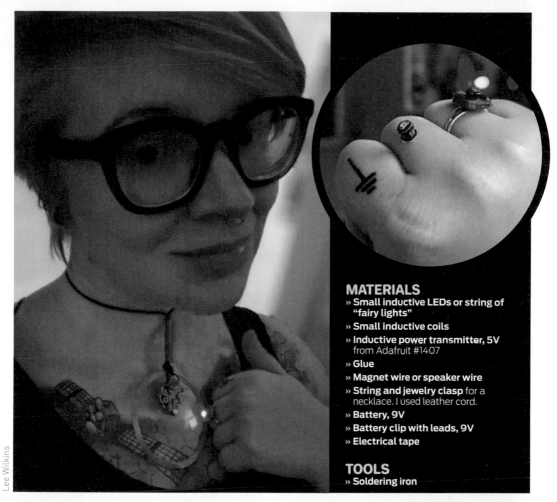

Lee Wilkins

MATERIALS

» **Small inductive LEDs or string of "fairy lights"**
» **Small inductive coils**
» **Inductive power transmitter, 5V** from Adafruit #1407
» **Glue**
» **Magnet wire or speaker wire**
» **String and jewelry clasp** for a necklace. I used leather cord.
» **Battery, 9V**
» **Battery clip with leads, 9V**
» **Electrical tape**

TOOLS
» **Soldering iron**

MAKE YOUR INDUCTIVE ADORNMENTS

Now that we have the basics down, how can we make this into some cool interactive adornments? In this project we'll make an inductive LED ring, then use the same technique to make fairy light sleeves. These will light up as your finger or hand passes over the necklace transmitter coil, so you can be illuminated as you dance! You can use this technique to make all kinds of inductive adornments.

1. MAKE THE TRANSMITTER NECKLACE
First we'll extend the leads on the coil to make a necklace. Clip the coil attached to the transmitter board (Figure **M**) and extend the two leads by soldering on much longer copper wire or speaker wire.

Twist both ends of the copper wire around

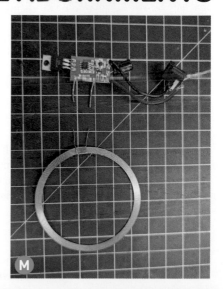

M

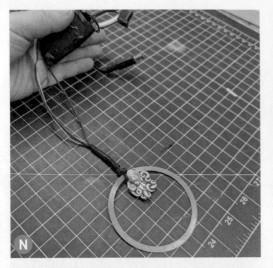

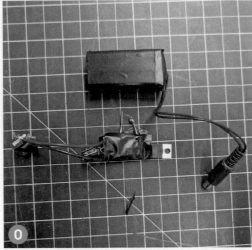

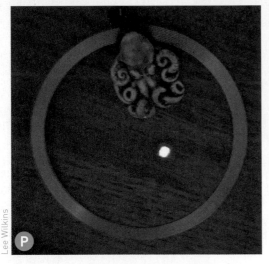

Lee Wilkins

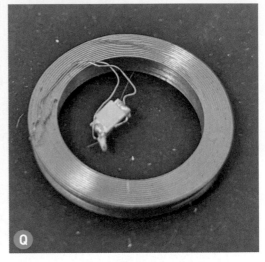

the string to make a necklace with the coil as a pendant. You can make it a choker, or a longer pendant depending on what you like! I added an existing pendant to make it fit my style a bit more; if you want a bit more inspiration, try modifying some jewelry you don't wear enough.

Once you're happy with your necklace, solder the other ends of the newly added wires back to the transmitter board. I also used some copper wire to wrap around the outside of the bundle of wires, just to hide the mess of the solder connections (Figure N).

2. POWER THE COIL

Solder the 9V battery clip onto the red and black wires of the transmitter board. Remember to use heat-shrink tubing or electrical tape to stop it from shorting. You can use hot glue to glue the board to the battery, or you can build another enclosure to hold it all together. I wrapped the board and battery in vinyl tape to make sure there were no shorts (Figure O).

3. TEST AND TROUBLESHOOT

You can test your circuit by plugging in the 9V battery and placing a small inductive LED inside the center of the coil (Figure P). You'll notice that as you rotate and move the LED, it might fade or turn off. You can get a feel for how the fields of the coils intersect. Try flipping your LED and coil backward as well!

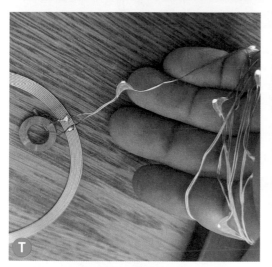

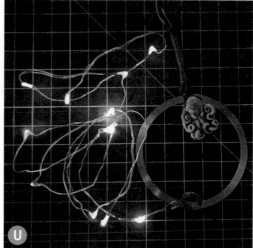

4. ADORN YOURSELF!

Now you can use the fairy lights or the small LEDs to create interesting interactions:

- **LED rings** I soldered a small inductive LED to a tiny receiver coil (Figure Q), then glued it to a blank ring base (Figure R). Bring it near the necklace coil (Figure S) to light it up!
- **LED sleeves** You can also solder string lights to a coil (Figures T and U), and glue this coil to a ring as well. Then you can light up a whole string of fairy lights. Wrap them around your hands or arms to make some cool light-up sleeves!

I'd love to see your ideas. Share your creations with me on Twitter @leeborg_ and let me know what you're doing with induction! ⊘

Smart and Sour

I created a sourdough monitoring system — and you can easily make one too

Written and photographed by Ivana Huckova

IVANA HUCKOVA is a software engineer at Grafana Labs, working on OSS Grafana. In her free time and during hack days, she loves to build fun monitoring solutions for uses like sourdough starters, candles, or avocado plants.

I have a confession to make: I'm one of those people who has been making homemade bread during the coronavirus lockdown. I even grew my own sourdough starter.

In case you didn't know, a starter is a yeasty mixture used to make bread rise. (Here's a basic recipe: kingarthurflour.com/recipes/sourdough-starter-recipe.) You make it out of flour and water, but it needs to be kept at the right temperature and regularly "fed" and partially discarded over the course of a week or two before it's activated and ready. Using it requires you to get it to just the right amount of rise before adding it to your recipe, but in the end it's definitely worth it.

I'm a software developer at Grafana Labs and also really interested in learning more about things that surround me, so when I heard it was possible to make a sourdough starter monitoring system, I had to try it. (Word is, it can improve the quality of the starter and the taste of your bread.)

Sourdough monitoring systems help you track your sourdough starter's temperature, humidity, and height. Mine uses an ESP32 development board, which collects data from sensors attached to the lid of the sourdough jar (Figure A). To measure the height of the sourdough starter, I have used an ultrasonic distance sensor. For temperature and humidity, I have used a DHT11 sensor. Produced data are then sent to the Prometheus database and they are visualized in beautiful graphs in Grafana. In Grafana, you can track your metrics in time or even set up alerts that will remind you when to feed your starter.

The system I made was inspired by sourd.io, which was created by Christine Sunu, and it's the first IoT project I've ever worked on. I enjoyed the process so much that I wanted to share it with everyone — especially anyone who'd like to dip their toe in IoT development. The setup is really easy, and I'm going to take you through the steps.

INSTALL SOFTWARE

First, install the **Arduino IDE** onto your computer. I used IDE version 1.8.10, as newer versions were throwing errors on MacOS Monterey. Version 1.8.10 was recommended as a solution on the Arduino forums and it worked for me.

Secondly, you will probably need to install **CP210x USB to UART Bridge VCP Driver** (silabs.

TIME REQUIRED: 2 Hours
DIFFICULTY: Easy
COST: $25

MATERIALS
» **ESP32-PICO-KIT V4** development board
» **DHT11** temperature/humidity sensor module
» **HC-SR04** ultrasonic distance sensor
» **HC-SR04 holders (2)**
» **Dupont cables, M–F (7)**
» **Micro-USB cable**
» **USB charger**
» **Jar, medium-sized** for your sourdough starter, with a lid that you can cut

TOOLS
» **Snips or rotary tool** or something to cut a hole in the middle of the jar lid
» **Glue and/or tape** such as a glue gun, super glue, or double-sided tape (ideally all of them)
» **Computer** with Arduino IDE (free from arduino.cc/en/software) and internet access

A

com/products/development-tools/software/usb-to-uart-bridge-vcp-drivers) if your OS won't recognise the USB Serial automatically. This driver basically lets your computer communicate with your ESP32 development board. Again, I am on OS Monterey and I had to install this driver.

SET UP A PROMETHEUS DATABASE

The next very important thing is to set up a database, where you are going to store the data from your sensors. In this project, we'll use Grafana Cloud, which comes with a free hosted Prometheus time-series database, and also Grafana for data visualization. To start, visit Grafana Cloud (grafana.com/auth/sign-up) to

sign up and create a new account. As soon as the account is all set up, you can see the portal with hosted Grafana and Prometheus instances (Figure **B**).

As soon as this is done, you'll need to click on the "Send Metrics" button next to the Prometheus card and create a new API key. Keep that API key and user information on hand, because you will use them soon to send your data to the database.

CONNECT SENSORS TO BOARD

DHT11

The DHT11 is a very basic and cheap sensor made of two parts — a humidity sensor and a thermistor. It can be used to measure the environment in which our sourdough starter lives.

If you have bought a DHT11 sensor module, then you should see three pins. No worries if you see four — you've probably bought just a sensor by itself. This has happened to me as well and it is an easily solvable problem — you just won't use the third pin (Figure **C**).

Connect these pins to the ESP32 board in the following way, as shown in Figure **D**:

- Vcc pin to 3.3V on board
- Data pin to pin 32 on board
- Gnd pin to Gnd on board

HC-SR04

The HC-SR04 ultrasonic sensor provides distance measurements from 2cm to 4 meters, with accuracy around 3mm. This sensor will be used to measure how far your sourdough starter is from the lid, and to calculate its rise and/or fall. The HC-SR04 has 4 pins that you will connect to the ESP32 board in the following way (Figures D and **E**):

- Vcc pin to 5V on board
- Trig pin (sends the signal) to pin 4 on board
- Echo pin (receives signal) to pin 5 on board
- Gnd pin to Gnd on board

PROGRAMMING

You next need to write and upload a program that will retrieve data from the sensors and send it to the Prometheus database. Connect your ESP32 board (with connected sensors) to your computer with a micro-USB cable.

SETTING UP ARDUINO IDE TO SUPPORT YOUR ESP32 BOARD

Open your Arduino IDE. If you are using an ESP32 board (rather than an Arduino board), you need to add a board definition that adds support for your ESP32. Go to Arduino → Preferences and input the url https://dl.espressif.com/dl/package_esp32_index.json to the "Additional Boards Manager URLs" input (Figure **F**). This open-source board definition adds support for programming ESP32 boards.

Then go to Tools → Boards → Boards Manager and install the ESP32 core (Figure **G**). Next,

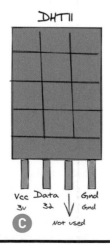

DHT11

Vcc Data   Gnd
3v   32   Gnd

C

Not used

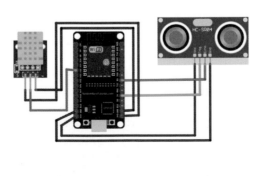

D

HC SR04

Vcc Trig Echo Gnd
5v   4   5   Gnd

E

GIVE A GIFT

One year of Make: magazine
for only $34.99 (regular newsstand price $59.96)

GIFT FROM:

NAME (please print)   42FGS1   **NAME** (please print)

GIFT TO:

ADDRESS/APT.   **ADDRESS/APT.**

CITY/STATE/ZIP   **CITY/STATE/ZIP**

COUNTRY   **COUNTRY**

EMAIL ADDRESS (required for order confirmation)   **EMAIL ADDRESS**

☐ Please send me my own subscription of Make: 1 year for $34.99

You can also subscribe online at makezine.com/give42

We'll send a card announcing your gift. Make: currently publishes 4 issues annually. Allow 4–6 weeks for delivery of your first issue. For Canada, add $9 US funds only. For orders outside the US and Canada, add $15 US funds only.

BUSINESS REPLY MAIL
FIRST-CLASS MAIL   PERMIT NO. 187   LINCOLNSHIRE IL

POSTAGE WILL BE PAID BY ADDRESSEE

Make:
PO BOX 566
LINCOLNSHIRE IL 60069-9968

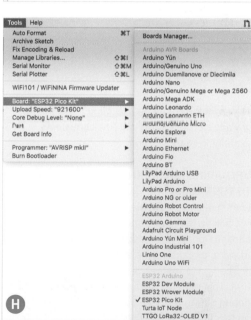

in Tools → Boards, choose the ESP32 Pico Kit board (Figure **H**). This tells the Arduino IDE which profile and base libraries to use when compiling the firmware image, and how to flash it to the board. Make sure that in the Tools → Port submenu, you have selected the new COM port.

ADDING LIBRARIES

Go to Sketch → Include Library → Library Manager (Figure **I** on the next page) and install the following libraries:

- DHT sensor library by Adafruit
- HCSR04 by Martin Sosic
- Adafruit Unified Sensor by Adafruit
- ArduinoBearSSL by Arduino
- ArduinoHttpClient by Arduino
- PrometheusArduino by Ed Welch
- PromLokiTransport by Ed Welch
- SnappyProto by Ed Welch

CREATING A PROGRAM

Download the Arduino sketch from github.com/ ivanahuckova/sourdough_monitoring_grafana. In the folder, you will find 2 important files: *sourdough_monitoring_grafana.ino* and *config.h*. In the first file, we have our program. In the second, we are going to store our variables.

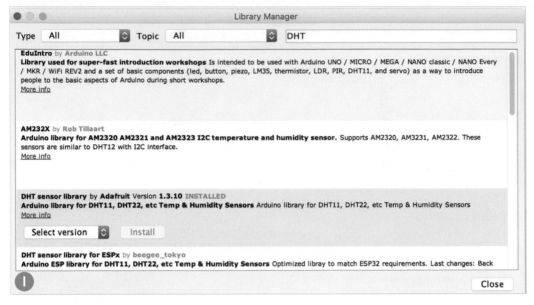

Open the *config.h* file and add your variables:
- Your Wi-Fi name and password, so the ESP32 can connect to your internet
- Your user ID as GC_PROM_USER
- Your API key as GC_PROM_PASS

Take a look through the *sourdough_monitoring_grafana.ino* sketch. You will find comments explaining each section.

UPLOADING AND RUNNING THE PROGRAM TO YOUR ESP32

Let's make this all come together!

Click on the open button (arrow pointing up), choose Open... (Figure **J**), find your *sourdough_monitor_grafana.ino* file, and open it.

Then go to Tools → Serial Monitor and open that. This monitor window will allow us to see the debug output from the board. Set the port speed on the bottom to 115200.

And lastly, use the IDE window's Upload button (the arrow pointing to right in the top bar of the IDE window) to build and upload the program to the ESP32 board.

Once the upload has completed, your Serial Monitor will show:
- Start-up and Wi-Fi connection
- Results of submitting metrics

If everything was set up correctly, you will soon see the incoming data also in your Prometheus database.

VISUALIZING YOUR DATA

Now that you have all your data flowing to Prometheus, it is time to visualize them in Grafana. To open Grafana just click the "Log In" button on your Grafana Cloud page (Figure **K**).

Next, we will make some dashboards. Just click on "Create Dashboard," add a new panel, and choose your Prometheus data source (Figure ). If you are not familiar with Prometheus query language, you can click on the Metrics browser and choose the metrics you want.

BUILDING YOUR SOURDOUGH JAR

Now comes the hands-on part. This is how I mounted the electronics in my jar, but I'm sure there is a much better way, so feel free to improvise with what you have.

Take the lid from the jar and carefully make two holes: one in the middle of the lid for the ultrasonic sensor and another, smaller one next to it for the DHT11.

Put the ultrasonic sensor in the HC-SR04 holder and glue it to the top of the lid. With double-sided tape, stick the DHT11 to the lid so it is facing down into the jar (Figure **M**).

Place the ESP32 board into the top of the HC-SR04 holder (Figure **N**). You can also carefully glue the board to the HC-SR04 holder with a glue gun. Make sure that you are only gluing to extended standoffs, not the board itself.

It should now look like Figure **O**. Just place it in the kitchen (or wherever you keep your sourdough starter) and plug it into the USB charger. I added a cover from a cut-plastic bottle to make sure the sourdough does not dry out through the holes in the lid.

READING THE GRAPHS

Grafana data is going to tell you more about the environment in which your sourdough starter is growing and about the starter itself.

In your graphs, you can see the temperature and humidity inside the jar, which might be actually different from conditions in the room. The warmer it is, the more often you have to feed your sourdough.

Moreover, I use the graphs to evaluate the height of the starter; when it has doubled in size and then starts to shrink, I know it's feeding time (Figures **P** and **Q**). I even set up alerts in Grafana that notify me when this happens.

I really hope you will try this out and let me know how it went! Also, if you have some fun and interesting IoT ideas, I would absolutely love to hear about them! Just tweet to me at @ivanahuckova. ✪

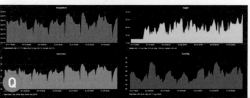

Zoom of Doom

Build a spooky, real-time video head-in-a-jar for your haunt

Written and photographed by Kevin Webb

KEVIN WEBB is an angel investor in startups that benefit biodiversity. In his spare time, he enjoys getting outdoors, visiting natural history museums, and making things. You can find him on Twitter @webbkevin or at his project site ktfoundry.com.

TIME REQUIRED: 2–3 Hours

DIFFICULTY: Easy

COST: $40–$400

MATERIALS

» **Projector** I used an Optoma EH336 ($150 on eBay), but even the Kodak Luma pico projector we tried ($300) worked in a brightly-lit room.
» **Laptop computers (2) with Zoom video call software** free from zoom.us/support/download. Use your highest-quality webcam for the greenscreen room.
» **Green foam board** from Staples, Michaels, etc., for the greenscreen. Cardboard or plywood painted bright green should work too.
» **Clamps** or other way to mount greenscreen perpendicularly to a table or desk
» **White plastic face** such as Amazon B0044S91IG
» **Large jar, plastic or glass** such as this 128oz one, Amazon B01LZFR2IF
» **Wire, bendable but stiff** I used a spare coat hanger.
» **Decorations (optional):**
  • **Wood or foam** for building a top and bottom structure
  • **Aluminum tape** for covering these structures
  • **Lights, LED or fiber optic or EL wire** I used Lightkraft's Chasing EL wire lights from etsy.com/shop/Lightkraft. They're battery operated and allow you to change the speed and direction of how the lights appear to move.

TOOLS

» **Box cutter or X-Acto knife**
» **Scissors**
» **Hot glue gun**
» **Drill**
» **Laser cutter/CNC machine (optional)**
» **Wood glue (optional)**

Ⓐ

Last Halloween my girlfriend and I hosted a Witches vs. Mad Scientists party (her excellent idea) for our vaccinated friends. As we brainstormed, I found my way to Mr. Chicken's website and was inspired by his Sybil the Clairvoyant crystal ball project (www.chickenprops.com/p/sybil.html). I wanted to see if I could create something similar, with three big differences: 1) use easy-to-find materials, 2) make it a floating head in a jar, à la *Futurama* (futurama.fandom.com/wiki/Heads_in_Jars), and 3) make the projection real-time, so friends could put their own heads in glass at the party.

Depending on what materials you have on hand, this project was surprisingly effective for how inexpensive it was! Here's how to do it.

1. MOUNT THE HEAD

To start, you'll want to suspend the head in the jar to create the illusion of floating. I glued the plastic head to a bent wire, which is run through a hole drilled in the lid. The mask I bought had a large lip that I trimmed off with scissors. Then I unfolded a coat hanger and bent the bottom in a big U shape that I hot glued around the inside of the mask. Near the center of the mask, bend your wire outward 1"–2", then add a 90° bend so the wire runs straight up a couple inches behind the mask.

Insert the mask and wire into the jar and figure out approximately where you want it to float. Drill a hole in the lid where the wire should pass through, then pass it through. Adjust the wire up and down until you determine the ideal height for the mask, then bend the wire 90° just above the lid. Add hot glue and make small adjustments to the wire to get the mask as close to centered as possible. Screw the lid on (Figure Ⓐ).

2. TEST THE PROJECTOR

Now let's see if the effect generally works. Get out your projector, set it a little back from the face, and on your computer, find a centered image of a creepy face. You can use any still image of a face, ideally with a black background. Adjust size, focus, projector angle, distance from the jar, and heights of jar and projector until it looks about right (Figure Ⓑ on the following page). A face should convincingly be projected on your head in a jar!

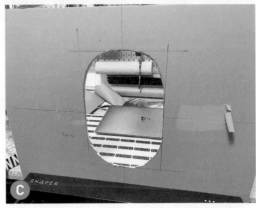

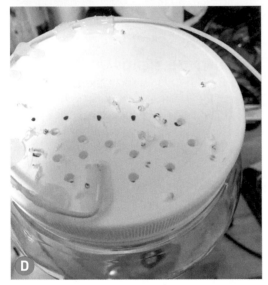

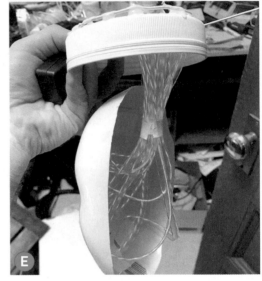

3. MAKE THE GREENSCREEN

To really create a Zoom of Doom, you need to make a greenscreen in another room with a hole cut in it that's big enough for your friends' heads. I used a $6 green foamcore board from Staples, then cut a hole big enough to fit my fairly large head (Figure C). I mounted the board perpendicular to a standing-height table using clamps I had in my office, then set a laptop on books so that its camera matched face height, and experimented with distances.

4. ZOOM OF DOOM!

Open your Zoom settings, choose Background & Filters, and check the box "I have a greenscreen." Find a big JPEG of nothing but the color black to serve as a background, then click the + button to

add it as a Virtual Background.

If you set up a Zoom conversation between your laptops, you'll now be able to project a friend's head in the jar, in real time. And if your sound is on, your guests can have conversations with your suddenly bodyless friend!

5. GET FANCY

You can now decorate your head in a jar however you see fit. For mine, I wanted metal top and bottom covers (again inspired by *Futurama*), and I wanted moving lights in the back of the head to make it seem as though it was being supported through high-tech tubes.

To get the light effect, I calculated how many lengths of EL wire (N) I wanted to string between the jar lid and the head, and I drilled ($N/2 + 1$)

holes, surrounding the hole where the coat hanger wire poked through (Figure **D**).

Through these I ran the strands of EL wire. When I had many loops of EL wire hanging from the lid at lengths/placements I liked, I then hot-glued the bottom arc of each wire to the back of the mask in different places. Tying the cables together with tape produced a nice, clean look (Figure **E**).

To make the "metal" covers, I used my laser cutter to create a wooden structure that would surround the jar at top and bottom, though you could do this nearly as easily by hand cutting foamcore or cardboard. After wood-gluing it together, I covered both top and bottom with aluminum tape, using layers to cover all exposed wood, including any that was visible through the bottom of the jar (Figure **F**).

HEAD-IN-A-JAR PARTY!

Setting this up at a Halloween party is a blast. As a final touch (which I didn't get to), I suggest finding a box or some cleverer way to hide the projector in front of the head. Since this effect works from all angles, placing your head in a jar on a table or something else guests can walk around works well. You may also want to hide the computer screen, or keep it visible so that the friend whose head is projected can see and hear whomever is talking with them.

GOING FURTHER

And that's it! If you're interested on building on the idea, the same general framework should work lots of different use cases. One that occurred to me would be to go way bigger: with a strong projector, you could turn your friends into Zordon from *Power Rangers*! At the other extreme, imagine creating shrunken heads in jars, or projecting a friend's face onto a creepy mannequin. On second thought, don't; a talking mannequin is entirely too scary.

If you're interested in the history of this effect, I went down several rabbit holes on "Madame Leota" from Disneyland's Haunted Mansion (disney.fandom.com/wiki/Madame_ Leota). I suggest checking out the great mini-documentary about the whole ride in *Behind the Attraction* on Disney+.

Servos and Micro:bit

It's easy to connect servomotors and program them in MakeCode!

Written by
Kathy Ceceri

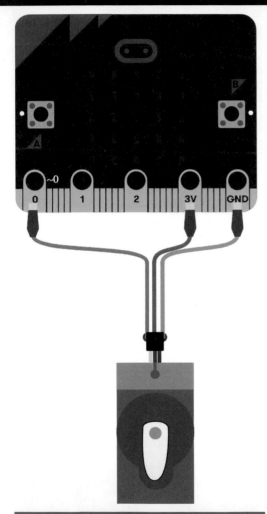

KATHY CECERI is an award-winning writer, educator, and maker with a focus on science, technology, history, and art. She develops teaching and learning materials for *Make:*, Adafruit Industries, and Girl Scouts of the USA.

In *Make:* Volume 79, Mario the Maker Magician showed how to code an Arduino microcontroller board to move a servomotor for a robot. It's easy. But here's an even easier way, using the micro:bit computer board and the popular MakeCode visual programming language!

SERVO BASICS

A *servomotor* is different from regular motors because you can control how far and how fast it turns using computer programming. Mini servomotors are great for using with simple robots, because they can be controlled directly by small boards like the BBC micro:bit. They're not very powerful, but they're perfect for lightweight designs made with paper and cardboard. (Bigger servos and regular motors need add-on hardware to work with microcontrollers.)

Just like regular motors, servos have a *shaft* — the part of the motor that sticks out and spins. To help you attach things to it, the servos you will use come with interchangeable *horns* — little plastic arms that snap onto the motor shaft. You usually get a variety of shapes with each servo. They all have tiny holes you can tie or hook things to, and come with a screw if you want to connect something to the servo more permanently.

There are two kinds of servos you may run into when building simple robots. For the projects in my book (which this article is excerpted from, see page 91), you will be using *positional* servomotors. These motors can only turn halfway around, then pivot back. In robotics, they're used for heads, arms, legs, and other parts that need to swing back and forth. For spinning wheels or cranks, small robots use *continuous* servomotors, which look just like regular servos but can rotate all the way around. Make sure you are using the right kind of servomotor (and the right MakeCode blocks) for your robot!

CONNECTING SERVOS TO MICRO:BIT

To connect a servo to a board like micro:bit, you use its cable. The cable is made up of three wires in different colors (Figure Ⓐ):

- **Orange or red is power.** It draws electricity from the micro:bit (or other source, such as a separate battery pack) to make the motor run.
- **Brown or black is ground.** It completes the circuit by bringing it back to the micro:bit.
- **Yellow or white is signal.** It carries the programming instructions from the micro:bit to the servo.

A connector at the end lets you plug wires into the servo cable. For the projects in this book, it's handy to use a connecting *jumper wire* of the style known as *alligator-clip-to-male-header-pin*. On one end, it has a pin that can plug into the servo cable. On the other end, it has an alligator clip that can clamp onto the edge connector rings on the micro:bit. (To open the alligator clip, squeeze on its "head" and the "jaws" will open up. Let go, and they snap closed. The "teeth" will usually give you a good grip on the part you are connecting to.)

It's very important to connect the wires from the servo to the correct pin on the micro:bit! You should also get used to attaching them in the correct order. This will avoid damage to the servo or the micro:bit. If possible, use alligator clip wires that more-or-less match the colors of the servo cable wires. If not, attach little labels to each wire with tape to keep them straight. Here is how to attach them (Figure Ⓑ):

- **First:** Brown or black (ground) connects to the GND pin on the micro:bit.
- **Second:** Yellow or white (signal) gets connected to one of the micro:bit's three programmable pins, labeled as 0, 1, and 2. For the projects in my book, you'll connect the servo to Pin 0.
- **Third:** Orange or red (power) goes to the 3V pin. The name stands for "3 volts," which is the amount of power the micro:bit can send to another device.

TIME REQUIRED: 30–60 Minutes
DIFFICULTY: Easy
COST: $30–$40

MATERIALS

» **micro:bit V2 microcontroller** V1 will also work, but won't include sound.
» **USB micro-B data cable** that fits your computer
» **Micro servo, 9g (positional, not continuous)** with servo horns that snap onto the shaft, and a cable with a plug that takes male jumper wires
» **Jumper wires, alligator clip to male header pin, in different colors (3)** to match the servo wires, preferably red, black or brown, and yellow or orange
» **Optional:**
  • **Extra-long USB data cable (3 feet or more)** so your bot can move around while connected to the computer for power and updating the program
  • **Additional male-to-female jumper wires** for longer distance remote control
  • **Battery pack for the micro:bit**
    ▪ The punch-out cardboard battery pack holder that comes with some micro:bits is handy for holding the board and batteries together; print and cut out your own from the micro:bit site (microbit.org/get-started/user-guide/battery-pack-holder).
    ▪ You can also get a larger pack that holds two AA batteries and has an on/off switch.

Ⓐ

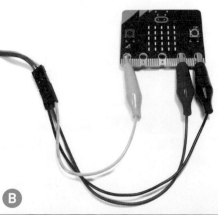

Ⓑ

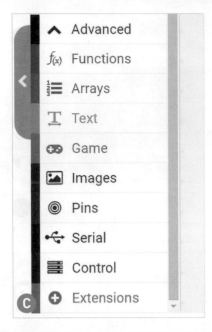

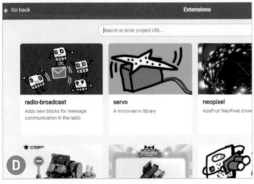

SERVOS AND MAKECODE

MakeCode is a free block-based programming language created by Microsoft to help students learn coding. It looks and feels very much like MIT's Scratch. The biggest difference is that your program runs on an *on-screen simulation* as you build it. You can program lights to flash when you push a button, or music to play when the board shakes, and watch the animated simulation to see if you got everything right. You don't need to download MakeCode to your computer. Build your program in your web browser, then download it to the micro:bit board. There's also a MakeCode Offline App that works on computers running Windows or MacOS.

To get started programming in MakeCode, check out the MakeCode micro:bit website (makecode.microbit.org). There are tutorials and sample projects on the main page to help you get started!

TO PROGRAM A SERVOMOTOR WITH MAKECODE, FOLLOW THESE STEPS:

1. MakeCode has special blocks to use with servos, but you have to add them to the list of categories. To find them:
   - Go to the bottom of the category list and select Advanced. (Figure **C**)
   - Scroll down to the bottom and click on Extensions.
   - You'll jump to a new page, with many different extras you can add to MakeCode. Look for "servo," with a green-and-blue drawing of a servo (Figure **D**). Click on it and you will jump back to the MakeCode workpage. The menu for Servos in dark green will appear in the middle of the other categories.

2. Open the Servos menu and drag the **set servo [P0] to [90]°** block to the workspace. Place it inside a block like **on start** to activate it (Figure **E**).

3. Click on the number 90 to open a slider that lets you rotate the servo horn to anywhere within a half-circle (Figure **F**). Positions are measured using a scale of zero to 180 *degrees*. (The teensy circle next to the number 90 is the

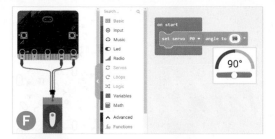

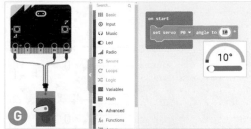

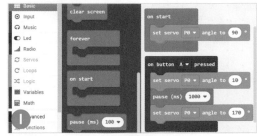

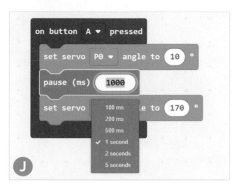

symbol for degrees. In math, a circle is divided up into 360 slices called degrees. They're like the minute marks on a clock face. Half a circle has half of 360 degrees, or 180.) The midpoint is 90 degrees, where the servo shaft points forward on the servo. This is the servo's *neutral position*, which means the motor is at rest. So always start and end with your servo at 90 degrees.

4. To move the servo to a different position, add more **servo** blocks to the stack, and change the position of the servo horn by using the slider or typing in the number of degrees. To avoid straining the little servomotor, don't make it go all the way to the ends. Limit its movement so it only goes between around 10 degrees and 170 degrees (Figures G and H).

5. To give the servo time to move before the program goes on to the next step, put a **pause** from the Basic menu (blue) after every **servo** block (Figure I). Set the pause for around one second; that's 1,000ms or milliseconds (Figure J).

6. Remember to end with the servo at 90 degrees (Figure K). That's it! ◙

Sarah Mather
and the
Aquascope

Build an underwater viewing scope to inspect boats, find fish — or spot enemy submarines

Written and photographed
by William Gurstelle

WILLIAM GURSTELLE's book series *Remaking History*, based on his *Make:* column of the same name, is available in the Maker Shed, makershed.com.

Adobe Stock - Terriana

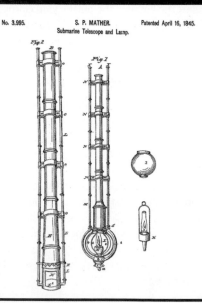

No. 3,995. **S. P. MATHER.** Patented April 16, 1845.
Submarine Telescope and Lamp.

Fig. 2 *Fig. 1*

A

On January 24, 1843, the editors of the Brooklyn *Evening Star* were caught up in whirl of excitement about an invention created by one of their own readers, Mrs. Sarah P. Mather. "The world," exclaimed the newspaper article, "is indebted to this inventor, who is not only *American*, but an *American Lady!*" A bit further on, the writer waxes even more effusively: "The fair inventress says to old Neptune and is obeyed: *Ground your Trident* old despot of three fifths of our World! Lay open your dark, hidden dominions!" (All of the italic words and exclamation points were in the original article.)

Wow! The underwater telescope that Mrs. Mather invented — better known today as an *aquascope* — certainly caught the attention of the New York press. Now, whether it truly merits that sort of acclaim or not lies in the eye of the evaluator, but after building a replica, I found Mather's aquascope an easy device to make and a fun one to use. The aquascope permits people on boats, docks, or piers to look beneath the surface of the water to get a better view of what is going on below. According to Mather's patent application, the underwater telescope "can be used for various purposes, such as the examination of the hulls of vessels, to examine or discover objects under water, for fishing, blasting rocks to clear channels" and so on.

As John Adie, Esq., wrote in an 1850 magazine article about the then-newfangled underwater telescopes, "The reason why we so seldom see the bottom of the sea or a lake is due to the irregular refractions given to the rays in passing out of the water into the air, caused by the constant ripple of the water where refraction takes place. Mather's aquascope punches through the optically unfriendly water surface, bypassing glare or surface refraction and allowing a clear picture of what goes on below.

Mather's patent application shows how her aquascope had a pair of unique innovations over previous scopes (Figure **A**). First, Mather outfitted her device with a lamp that could be lowered into the water along with the aquascope. This made it possible for the device to be used at night. Second, Mather's aquascope incorporated an angled mirror that gave the device the ability to look not just down but also sideways in the water, which was very helpful for inspecting ship hulls, piers, and so forth.

When Mather lived, it was unusual for a woman to obtain a patent on an invention and then attempt to market the device. But this she did, and that's about all we know of Sarah Mather. The records are sketchy, but we do know she was born Sarah Porter Stinson or Stimson around 1796 and died in New York City in 1868. In 1819 she married Harlow Mather, a distant cousin of the famous Puritan clergyman Cotton Mather, and they had several daughters, including Olive M. Devoe who in 1868 petitioned Congress for money to test her invention of a "submarine illuminator," building on her mother's work. When not inventing, Mrs. Mather was involved in the publishing of poetry, and near the end of her life she raised money for the Union Home and School for New York children of soldiers killed in the Civil War. Olive was the director and principal.

THE AQUASCOPE OR UNDERWATER TELESCOPE

This project is fairly simple, and you can change the dimensions to suit your needs. This aquascope uses a 3-inch-diameter, 2½-foot-long viewing tube, but you can change the length and diameter if you want a wider or deeper view. Because the air inside the scope makes it buoyant, you may want to add weight to the bottom to make it easier to push it down into the water.

At the bottom end of the tube is a rotatable mirror which can be adjusted to provide a straight-through view to the bottom, a 90-degree view to the side, or anything in between.

MAKING THE MATHER AQUASCOPE

Before you assemble anything, cut these pieces to the sizes described in the materials list: the PVC tube, the clear plastic tube, the round piece of wood, and the round piece of acrylic plastic.

1. With a saw, cut a notch for your nose in the top of the 3" PVC tube (Figure **B**). Smooth the surface with sandpaper because you will press your face against the tube when you use it.

2. Use the electric drill to drill two ⅜"-diameter

holes across from one another in the center of the clear plastic tube, 2" from bottom. Once the first hole is drilled, insert the ⅜" bolt and mark the spot directly across from the first hole with a marker, then drill the second hole. The holes should line up directly across the tube from one another — in other words, they should be diametrically opposed — and the bolt should be perpendicular to the interior surface of the tube.

TIP: Wrap a piece of masking tape around the tube before you drill, and mark the spot you wish to drill with a marker. This will make it easier to start and end drilling the hole without slipping.

3. Glue the mirror to the wood backing piece.

4. Insert the bolt into the plastic loop clamps. Fasten the clamps to the wood backing piece with the wood screws. When you insert the bolt and tighten the screws, the clamps should hold the bolt securely. If the bolt rotates, remove the screws and wrap it in tape so the clamps have a better grip. Next, remove the bolt from the clamps.

5. Place a washer on the bolt, and then an O-ring. Now, thread the bolt through one hole in the clear plastic tube, then through the loop clamps, then through the opposite hole in the clear tube. Attach the second O-ring, the second washer, and the wing nut (Figures **C** and **D**). Using the wrench, tighten the nut to seal the O-rings against the outside of the clear tube.

C

D

E

F

6. Once the mirror is in place, use the silicone sealant to glue the 3" clear acrylic disk to the bottom end of the clear acrylic tube (Figure **E**). I tied mine on tight until the sealant dried.

7. Check the fit between the clear plastic tube and the interior of the PVC tube. Apply a layer or two of duct tape to the upper edge of the clear tube if the fit is not tight.

   Apply silicone adhesive/sealant to the upper 1" of the clear tube (or its duct tape) and then insert it 3" deep into end of the PVC tube. I also ran a bead of sealant all around the joint (Figure **F**). Let the silicone cure according to label directions. Before the sealant dries, rotate the clear tube so the mirror points 180 degrees opposite the nose notch in the PVC tube (or the angle of your choice).

8. Use the silicone to attach the waterproof light to the side of the PVC tube so that it shines in the direction that the mirror faces.

9. Let all the glue and sealant joints dry. With your wrench, adjust the angle of the mirror to 45 degrees to look sideways.

10. Test the aquascope for leaks. If it leaks, apply additional sealant, and check the fit of the O-rings.

11. Stick the LED light to the side of the scope, pointing the direction your mirror is looking.

USING THE AQUASCOPE

Like Mather's original aquascope, this project allows the user to make nighttime observations and to look 90 degrees to the side of scope. But it's also adjustable so you can look in other directions.

- To look 90 degrees to the side, keep the mirror at a 45-degree angle.
- To look straight down, position the mirror vertically, so you can see past it, straight down through the clear bottom disc, and then tighten the bolts.
- Whichever angle you look, you could add more LED lights pointed in that direction. ●

Deep Space Transformer

Print this James Webb Space Telescope model and unfold it like the real thing

Written and photographed by Brian Mernoff

TIME REQUIRED:
1–2 Hours + 2–3 Days Print Time

DIFFICULTY:
Intermediate

COST:
$25–$35

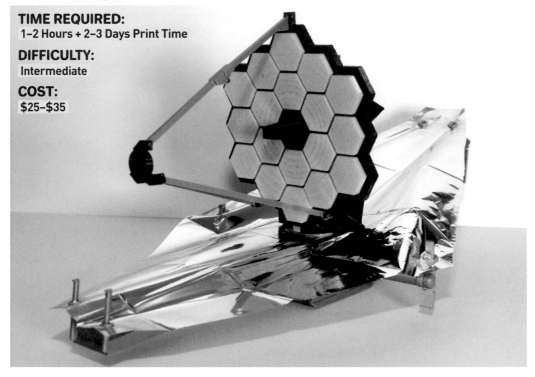

When making a new model, I look for two things. First, the subject must be something I'm excited to learn about. Second, it needs to present a design challenge: a new mechanism, a complex shape I don't know how to model, or a new tool or technique I haven't tried before.

The James Webb Space Telescope (JWST) was the perfect combination. I've always been a huge space fan and I had followed the progress of JWST for years. After the launch, I tracked the

telescope deployment steps online at jwst.nasa.gov. I'm a huge *Transformers* fan too, and what I was watching was pretty much a real live transformer! I had to figure out how to make a model that transformed in the same way.

My process begins with a design phase in my head. I researched schematics of the telescope and detailed videos and websites describing the deployment steps. During my hour-long commute I thought of different mechanisms and

BRIAN MERNOFF is the education coordinator at the MIT Museum in Cambridge, Massachusetts, where he explores how to use current research to engage visitors with the processes and tools of STEAM.

ways to create new hinges, move the secondary mirror assembly, and deploy the sunshield.

Then I started CAD design in Fusion 360. First, I determine the overall size of the model. In this case, the limiting structures were the hinges because the very large sunshield parts needed to fit on my Prusa MK3S printer. I compromised by scaling the hinges so I could use tiny nails, instead of filament or printed parts, as the axles.

Next I imported a schematic into Fusion 360, scaled it to match, and created the basic outlines of the parts. Then I worked through the model building the mechanisms. The secondary mirror assembly required a lot of testing by moving the parts and adjusting the angles.

A bigger challenge: The left and right boom arms for the sunshield needed to occupy the same space when the telescope was folded. I looked at umbrellas and toy lightsabers, and settled on making one boom smaller so it could store itself inside the other (Figure A).

After I printed and test-fit the parts, I added small magnets to hold the primary mirror wings tight (Figure B), as the side hinges turned out to be too loose.

The sunshield was by far the most challenging problem. I quickly found that mylar begins to tear whenever it is cut or punctured. Looking at a tarp, I remembered that the metal grommets around the holes are there to prevent the propagation of tears. I printed some grommets and glued them on the mylar, and it worked!

The final model measures 40cm (15¾") long and I'm very happy with how it came out. I can follow along with all the deployment steps and reverse them (Figures C, D, E, and F).

You can find this and other models I've designed on Thingiverse, Printables, and MyMiniFactory, username *chemteacher628*. I enjoy contributing to the 3D printing community since I've learned so much from the open source resources that so many people create. I hope you'll build this JWST model and learn as much as I did, and that maybe it will encourage others to explore science and engineering. ◉

See more of this project at makezine.com/ go/3DP-JWST, and check out Prusa's build video at youtu.be/N_Q_YUDUSLs.

MATERIALS
» **PLA filament in silver, black, blue, and gold** Download the free 3D files for printing at printables.com/model/109816 and follow the assembly instructions there.
» **Magnets, round, 6.3mm×2mm (4)**
» **Mylar sheet, aluminized** sold as a gardening product or as cheap thermal blankets
» **Glue**
» **Small wood nails, 1.25mm diameter**

TOOLS
» **3D printer**
» **Tinsnips or Dremel** for cutting nails
» **Drill bit, ¹⁄₁₆"** to clean up hinge holes

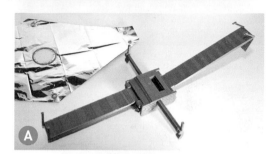

A

B

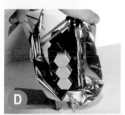

C

D

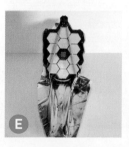

E

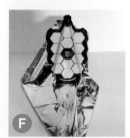

F

The Arduino ARM Family

Classic Arduinos use AVR processors, newer boards use ARM — so which is right for you?

Written by Michael Shiloh and Massimo Banzi

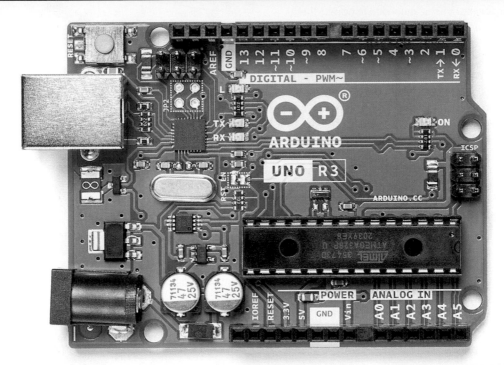

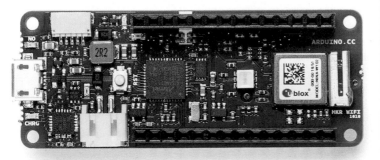

This article was adapted from the bestselling, newly revised *Getting Started with Arduino, 4th Edition*, available at Maker Shed (makershed.com) and fine booksellers.

The original Arduino family of boards was based on Atmel AVR 8-bit microcontrollers. They're excellent in terms of price, flexibility, and ease of use, but their limited processing speed and small memory size make it difficult to support modern networking protocols like Bluetooth, Wi-Fi, and GSM cellular. So Arduino has taken advantage of the availability of low-cost 32-bit microcontrollers based on the ARM architecture to create a family of dramatically more powerful and flexible boards.

WHAT'S THE DIFFERENCE?

Both AVR and ARM refer to families of microprocessors. The ARM architecture was developed by the ARM company and is licensed to other companies, while the AVR architecture was developed by Atmel and pretty much stayed within Atmel (now owned by Microchip).

The AVR-based microcontroller line started with relatively simple and slow 8-bit processors, and the product line has since grown to include 16- and 32-bit processors. Designed from the start to be the core of a microcontroller, the AVR processor has efficient commands for manipulating individual bits in input-output ports, while the more generic ARM processor might lack these features.

ARM-based microcontrollers, on the other hand, are typically 32-bit devices with more complex peripherals and substantially more memory, running at speeds greater than AVR devices.

WHY 32 BITS?

The phrases "8 bits," "32 bits," and "64 bits" are seen frequently, but what do they really mean? They mean the microcontroller's internal pathways can carry that many bits of data *at the same time*. So when a 32-bit microcontroller wants to get information from memory, it can get 4 times as much as an 8-bit microcontroller could in the same amount of time, just as a 32-lane highway could carry 4 times as many cars as an 8-lane highway. It also means that most of the internal processing, such as mathematical calculations, work on 32 bits at a time, so those calculations are much faster. This 32-bit *word size*, coupled with the faster clock speed, makes these boards practical for larger programs and more complex calculations where an

8-bit microcontroller might not be able to read sensors, analyze data, make decisions, and output control signals fast enough.

WHICH IS BETTER?

Depends what you're trying to do. Generally speaking, AVR-based systems will be less expensive and simpler to design and program. If you're just getting started, you almost certainly should start with AVR Arduino boards.

On the other hand, if you feel comfortable with Arduino circuits and programs, and need special features like wireless networking or complex mathematical calculations, then an ARM Arduino is probably more suitable because the increased word size, processing speed, and memory are better able to handle larger, faster, more complex programs.

THE ARDUINO ARM BOARDS

The ARM family of Arduino boards have used three variations of the ARM core: Cortex M0, M0+, and M4. The Cortex M0 core was optimized for low cost as a 32-bit replacement for 8-bit microcontrollers. The Cortex M0+ was further optimized to reduce power and add new features.

The Cortex M4 is a much more powerful core with a range of new features designed to support industries such as motor control, automotive, power management, embedded audio, and industrial automation with the addition of *DSP (digital signal processing)* instructions and optional *FPU (floating point unit)* that make mathematical operations extremely fast.

You can compare the current lineup of ARM Arduinos in the table on the following page.

SPECIAL FEATURES

Some of these boards have other special features:
- The Arduino MKR Zero has an SD card socket and I2S port, so it can play and analyze audio files and connect directly to *I2S digital audio* devices.
- The Arduino MKR Vidor 4000 adds a *field-programmable gate array (FPGA)* that basically allows you to design integrated circuits. Because your design is implemented in hardware, as opposed to software, a project implemented on an FPGA is incredibly fast. The

ARM FAMILY OF ARDUINO BOARDS

BOARD	FOOTPRINT	MICROCONTROLLER	RADIO PROTOCOL
ARDUINO ZERO	Uno R3	ARM Cortex-M0+	None
ARDUINO NANO 33 BLE, BLE SENSE	Nano	ARM Cortex-M4	BLE and Bluetooth
ARDUINO NANO 33 IOT	Nano	ARM Cortex-M0+	Wi-Fi, BLE, and Bluetooth
ARDUINO MKR ZERO	MKR	ARM Cortex-M0+	None
ARDUINO MKR WAN 1300, 1310	MKR	ARM Cortex-M0+	LoRa (low bandwidth, long range)
ARDUINO MKR VIDOR 4000	MKR	ARM Cortex-M0+	Wi-Fi, BLE, and Bluetooth
ARDUINO MKR NB 1500	MKR	ARM Cortex-M0+	Internet over 4G GSM network
ARDUINO MKR WIFI 1010	MKR	ARM Cortex-M0+	Wi-Fi, BLE, and Bluetooth
ARDUINO MKR GSM 1400	MKR	ARM Cortex-M0+	Internet over 3G GSM network

Vidor also adds a *micro-HDMI port* — because it's fast enough to generate video frames in real-time.

OPERATING VOLTAGE

In contrast to the Arduino Uno, which operates at 5 volts, the ARM boards operate at *3.3 volts*, so they will run off a single-cell rechargeable Li-Ion or Li-Po battery. Some, like the MKR WiFi 1010 and MKR WAN 1310, include a battery connector and charging circuitry that will charge the battery whenever USB power is available — perfect for wireless battery-operated projects.

Operating at 3.3V means that you must take this into consideration when connecting external components such as LEDs and sensors. Switches and resistive sensors will work fine, but active sensors designed for 5V, such as some temperature and humidity sensors, may not work properly at 3.3V. Extreme caution must be used when mixing both 3.3V and 5V components in a circuit. Voltages greater than 3.3V must never be present at any pin of a 3.3V component.

DRIVE CURRENT

Each pin on an AVR Arduino should be used for at most 20 milliamps. But for ARM Arduinos based on the SAMD21 microcontroller (those with the ARM Cortex-M0+ core) this number is only *7 milliamps!*

Assuming a worst-case LED voltage of 1.8V, 3.3V – 1.8V leaves 1.5V on the resistor, and solving Ohm's law for resistance $R = V / I = 1.5 / 0.007$ $= 220\Omega$. This means you should always use a resistor of at least 220 ohms with an LED. If your LED is too faint you'll need to use a transistor.

DIGITAL TO ANALOG CONVERTER

Although all Arduinos support the `analogWrite()` function, the AVR Arduinos simulate an analog voltage by using digital *pulse-width modulation (PWM)*. This works fine to control the brightness of LEDs and the speed of motors, but sometimes you might need a true analog voltage. In that case, the ARM-based boards are ideal because they all contain a *digital to analog converter*, or *DAC* (except the Nano 33 BLE boards). This device does exactly what you'd expect: You give it a number, and it generates a voltage proportional to that number. This can be invaluable for controlling a variety of devices.

USB HOST

Arduino boards based on the Cortex M0+ can configure a USB port in *Host mode*. This means that rather than being a dumb device subject to the whims of a USB Host, your Arduino can be a USB Host initiating transactions with other USB devices — or can pretend to be a USB keyboard or mouse to control, invoke, or send data to an attached computer.

NANO AND MKR FOOTPRINTS

Except for the Arduino Zero, with its traditional Uno R3 footprint, all other ARM-based Arduino boards are in the smaller Nano or MKR footprints. Apart from size, what sets these apart is the type of connector used for the pins: Instead of sockets on top of the board, the Nano and MKR footprints have pins on the bottom of the board, suitable to plug directly into a breadboard. Similarly, whereas shields sit on top of an Uno, shields for MKR or Nano boards sit underneath. ◉

How to Become an
Amateur
Scientist

Adventures from a life of DIY electronics, instruments, and experiments

Written by Forrest M. Mims III

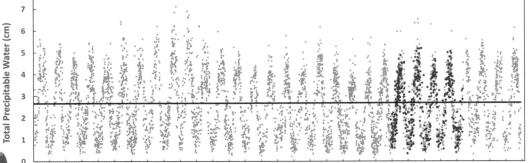

30 years of total water vapor measured by LED sun photometer (blue) and NOAA GPS (red).

Forrest Mims' 1970-75 workbench.

After *Make:* editor Keith Hammond saw my latest scientific paper, he suggested that I devote a column to how I do science without formal training. (My sole degree is a BA in government and English from Texas A&M.)

That paper is "A 30-Year Climatology (1990–2020) of Aerosol Optical Depth and Total Column Water Vapor and Ozone over Texas" (*Bulletin of the American Meteorological Society*, January 2022, doi.org/10.1175/BAMS-D-21-0010.1). The paper details 30 years of sun measurements from the field outside my rural office (Figure ). Included are seven charts speckled with thousands of data points that depict the recovery of the ozone layer over my Texas site since the historic volcanic eruption of Mount Pinatubo in 1991. Other charts depict a slight reduction in the atmosphere's aerosol optical depth (haze) and the absence of a trend in total column water vapor.

The reduction in haze, which is also obvious in thousands of sun and sky photos made during measurement sessions, is likely due to the closure of many coal-fired power plants. Since global warming models assume an increase in total water vapor over time, the absence of a trend in water vapor was surprising.

The accuracy of the LED sun photometer that made the measurements was supported by a nearby NOAA GPS receiver, and a NASA AERONET site in Oklahoma found a decline in water vapor over 24 years. Reductions in total water vapor across North China and the Central U.S. have also been reported.

Surprises like the absence of a long-term water vapor trend are what make science projects so fascinating. You never know what might be discovered. If you have an interest or experience in, say, culturing bacteria, or working with chemicals, building electronic or mechanical devices, you are already equipped with the basic skills to pursue a science project.

STUDENT EXPERIMENTS

My science adventures began while watching an oscillating fan during a seventh-grade class in 1957, when schools were not air conditioned. That fan suggested to me a method for controlling the flight of a rocket without using fins. Air would enter a forward-facing port in the nose cone

Forrest Mims and his many atmospheric instruments.

FORREST M. MIMS III is an amateur scientist and Rolex Award winner, was named by *Discover* magazine as one of the "50 Best Brains in Science." He has measured sunlight and the atmosphere since 1988. forrestmims.org

and be ejected from the side of the nose cone to provide course changes. While I was unequipped to implement this daydream design, it led to a series of career-changing projects.

During high school, I experimented with simple analog circuits using potentiometers to add numbers. This led to the idea that a pot is really a manually programmable memory device. That idea culminated in an analog computer that could be programmed to translate 20 words of various languages into English (makezine.com/go/homebrew-analog-computer). I donated it to the Smithsonian Institution after their mathematics curator visited to collect my early hobby computer documents. She described the language translator as an early example of hobby computing.

My great-grandfather unknowingly inspired another project. He was blinded by an explosion while working on a railroad bed in 1906. I'll never forget how he tapped his cane to hear echoes from power poles when we walked along a country road. That walk inspired the goal of developing a handheld travel aid for the blind when I was a senior at Texas A&M in May 1966. Texas Instruments had just developed powerful infrared (IR) LEDs, and they provided three of the expensive devices for my travel aid project.

B

Eyeglass travel aid for the blind (1971).

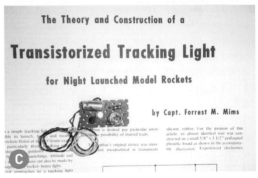

C

Light flasher article in Sept. 1969 *Model Rocketry* magazine.

I built a code practice oscillator circuit to pulse one of the LEDs. Infrared reflected from objects was detected by a silicon solar cell connected to an old hearing aid amplifier. The result was a miniature travel aid that emitted a hum from an earphone when objects were detected. I tested the travel aid with many blind children and adults in Texas and Vietnam. Eventually, I made a travel aid small enough to fit on sunglasses (Figure **B**). Unfortunately, I was unable to persuade a company to manufacture it due to liability concerns.

ROCKETS, LASERS, AND THE FIRST PERSONAL COMPUTER

So, I returned to my seventh-grade rocket idea, which I called *ram air control*. A class at the Air Force Academy did wind tunnel tests of the idea. Meanwhile, I launched test rockets at night and photographed their flame trails to determine how much control the ram air was exerting.

Recovering night-launched rockets required a flashing tracking light, which I made from the same circuit that pulsed the IR LED in the travel aid. The tracking light worked so well that I wrote an article about how to make it, "Transistorized Tracking Light for Night Launched Model Rockets" (makezine.com/go/rocket-flash), published by *Model Rocketry* magazine in September 1969 (Figure **C**). I had no idea the magazine would pay $93.50 for the article, which led to my decision to become a freelance writer after the Air Force.

I was then working with powerful lasers at the Air Force Weapons Laboratory in Albuquerque, New Mexico. My co-worker Ed Roberts and I often

discussed forming a company to sell electronic kits. When I showed Ed the light flasher article, he agreed it could become our first project. We recruited Bob Zaller and Stan Cagle and formed Micro Instrumentation and Telemetry Systems (MITS).

MITS sold only a few hundred light flashers and model rocket transmitters, so we moved to a series of other projects, including a diode laser I designed and the Opticom light-wave communicator Ed and I designed. *Popular Electronics* published the Opticom project as a cover story. When I left the Air Force and MITS in 1970 to become a full-time writer, I wrote manuals for calculators Ed designed at MITS. In 1974, Ed developed the Altair 8800 hobby computer kit. He gave me one of the first assembled Altairs in return for writing the manual. (See my column in *Make:* Volume 42, "The Kit That Launched the Tech Revolution," makezine.com/2015/01/08/the-kit-that-launched-the-tech-revolution.)

The Altair project became the cover article for the January 1975 issue of *Popular Electronics*. When Paul Allen and Bill Gates saw the magazine, they developed a version of BASIC for the Altair. Ed hired Allen to head the MITS software group in 1975, and Gates joined Allen at MITS that fall. A few months later they formed Microsoft.

WRITER, INVENTOR, ATMOSPHERIC SCIENTIST

Meanwhile, I moved away from the rocket project and developed optical fiber communicators, one of which I used to send data through an optical

Forrest Mims

D

Six-channel LED sun photometer being tested where early sun photometers were first tested behind the Smithsonian "Castle" in 1902.

E

TOPS-1 and TOPS-2 found a drift in NASA's ozone satellite in 1993.

fiber from a helium-inflated balloon to a ground receiver. In 1972, I discovered that LEDs can detect light at wavelengths just below their peak emission wavelength. I used this principle to develop a two-way communicator that sent data through air or an optical fiber with only two LEDs.

In 1989, *Scientific American* assigned me "The Amateur Scientist" column. Only three columns were published after the editor learned I reject Darwinian evolution. My termination became a national news story. It also motivated me to prove I could do publishable science. Therefore, I continued working on what would have become the fourth column, a 2-channel LED sun photometer that measured haze and total atmospheric water vapor. I tested the original and a 6-channel version behind the Smithsonian "Castle" in Washington, D.C., where such measurements were begun in 1902 (Figure D).

The optical filters used in conventional sun photometers are expensive, fragile, and degrade over time. LEDs are cheap, sturdy, and long lasting. My original LED sun photometer, which is still used today, played a key role in the 30-year paper mentioned above. After I published a formal paper on using LEDs in sun photometers,

they became used for this purpose around the world.

Meanwhile, I developed TOPS (Total Ozone Portable Spectrometer), a handheld instrument that measured the ozone layer (Figure E). TOPS would have been my fifth column in *Scientific American*.

In 1992, the American Scientific Affiliation

F The world standard ozone instrument, Dobson 83, during Forrest Mims' first visit to MLO in 1992.

asked me to give a talk about the *Scientific American* affair at a meeting at University of the Nations (UofN) in Hawaii. That trip provided the opportunity to calibrate my sun photometers and ozone instruments at NOAA's world-famous Mauna Loa Observatory (MLO). Earlier, my two TOPS had found a drift in ozone measurements by NASA's ozone satellite, Nimbus 7. During my first visit to MLO, the drift measured by the two TOPS was affirmed by the world standard ozone instrument (Dobson 83), which was being calibrated at MLO (Figure **F**). This finding became my first paper in *Nature*, the world's leading scientific journal.

The UofN talk also led to an assignment to teach an annual hands-on science short course, and that provided me with annual MLO calibration opportunities for 17 years.

In 1993, TOPS received a Rolex Award. This provided the money to hire my friend Scott Hagerup to develop Microtops, a microprocessor-controlled TOPS. When Solar Light acquired rights to Microtops, they developed Microtops II, hundreds of which are used today by scientists around the world to measure the ozone layer, water vapor, and haze.

In 2016, 24 years after Dobson 83 confirmed TOPS had found a drift in NASA's ozone satellite, NOAA hired me to calibrate Dobson 83 during a 64-day stay at MLO (Figure **G**). A comparison

G

Forrest Mims takes a break from calibrating Dobson 83 during a 64-day stay at MLO in 2016.

of 194 ozone measurements by Microtops II and Dobson 83 differed by only 1.9%.

While I continue to make regular measurements of the sun's irradiance, haze, and the total water vapor and ozone columns, my most exciting project is twilight photometry. My ultra-simple twilight instruments have twice been described in *Make:* and will soon be published in a scientific journal.

You can learn much more about my career as an amateur scientist, including my research for NASA during two Brazil campaigns, at www.forrestmims.org. The 30-year paper demonstrates the importance of persistence in doing long-term environmental research. On February 5, 2022, my sun and sky measurements reached 32 years and exceeded the longest series of column water vapor and optical depth measurements since the Smithsonian's at Table Mountain, California, from 1926 to 1957. (They also found no trend in total water vapor.) My goal is 50 years — if I live to 95.

YOU CAN DO THIS

I close with a challenge: Every issue of this remarkable magazine includes amazing projects by highly creative individuals who may be more equipped to do science than they realize. Besides making discoveries, doing science can be a fascinating hobby. I hope you will join me in doing science, even if you lack a science or engineering education. ❂

SIX HABITS OF A HIGHLY SUCCESSFUL DIY SCIENTIST

I've been fortunate to work with Forrest Mims for 15 years editing his column in *Make:*. Not only has he taught electronics to generations of beginners, he's made a name for himself as an atmospheric scientist too. In this overview of his career I recognize many of the habits of work, and of mind, that contributed to his success:

- **Follow your imagination.** (ram air control for rockets, seeing with sound)

- **Pursue novel theories to see where they lead.** (pots can be memory devices, LEDs can be sensors)

- **Investigate practical uses of your discoveries.** (IR rangefinder travel aid, rocket tracking lights, fiber optic communications)

- **Need an instrument that doesn't exist? Make it!** (TOPS ozone spectrometer, LED photometers)

- **Persist despite the skeptics.** (*Scientific American*)

- **Diligently collect data, rain or shine.** (32-year series, and counting!)

These are some of the steps Forrest followed to become an extraordinary amateur scientist, and you can too. Read more from Forrest online at makezine. com/author/forrest-m-mims-iii, in back issues of *Make:* (makershed.com/ collections/back-issues), and in his book *Make: Forrest Mims' Science Experiments*. And don't miss our DIY Science book series with illustrated guides to home chemistry, biology, astronomy, and forensic science, all available at makershed.com/books.
            —*Keith Hammond*

Making Fun

What if ... famous artists created these novelty toys?

Written and photographed by Bob Knetzger

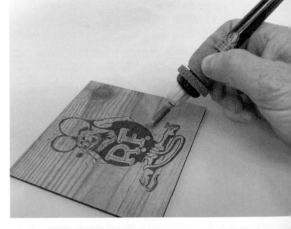

BOB KNETZGER is a designer/inventor/musician whose award-winning toys have been featured on *The Tonight Show*, *Nightline*, and *Good Morning America*. He is the author of *Make: Fun!*, available at makershed.com and fine bookstores.

Many famous artists have used their talents to create toys and products just for kids. Surrealist Salvador Dalí designed the colorful Chupa Chups lollipop logo. Modern sculptor Alexander Calder launched his career by creating an entire toy circus from wire and wood, and he later designed toddler pull-toys for Gould Manufacturing in the 1920s. Car designer Ed Roth and husband-and-wife architect/designers Charles and Ray Eames all created for kids.

But here are some fun "fakes" they *didn't* design. These plausible parodies never really were, but what if ... ?

Alexander Calder was trained as an engineer, but loved art and created his own toys from a young age. His unpretentious approach featured common materials and improvised fabrications using everyday items like wire, tubing, and springs. He called his playful hanging sculptures "mobiles," which innovatively used color and motion — and so does this magnetic spinning toy (Figure )!

To make this fun fake I hacked an existing Whee-Lo toy and added my own bent coat hanger wire handle in an elephant shape as a nod to Calder's own animal drawings and his famous bent-wire *Circus*. I used a Glowforge laser cutter to cut and engrave acrylic to add a "Calder" signature logo badge that joins the new handle to the old toy (Figure 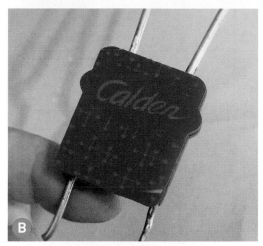).

Ed "Big Daddy" Roth was a mid-century maker of all things monster. He introduced the iconic iconoclast Ratfink, airbrushed monster T-shirts, modernized the hot rod, and made millions of monster/model car kits. Tolerated by moms and loved by kids, Roth's greatest creation was probably his own over-the-top-hat counterculture character. This woodburning set could have been a hot toy in its time — had it ever existed (Figure ).

A flea-market-find vintage woodburning kit provided the plug-in iron, instructions, and colored hot foils. The Glowforge's built-in camera made quick work of burning the Roth artwork into some sample wooden project panels. I even found an old photo of Roth using a woodburner iron to work on an acrylic bubble top car — the perfect

D

shot for the fake box (Figure **D**)!

Designers **Charles and Ray Eames** pioneered the forming of flat, laminated wood into curved shapes to make chairs and furnishings. They "took their fun seriously" and created witty kid's furniture and clever toys. Maybe they could have designed some skateboard decks — but not these mini-sized fingerboards (Figure **E**). Still, it's fun to imagine.

The angled kicktail mini deck is the perfect parody of Eames' trademark bent wood technique (Figure **F**). The graphic designs are laser etched into thin plywood. The trucks and wheels were cannibalized from existing toy fingerboards.

Antonio Gaudí is Spain's most celebrated architect, known for his modernist style. He created entire languages of shapes and forms, which he used in his designs for his Sagrada Familia cathedral, classic Art Nouveau residences like Casa Mila, and the colorful and playful gardens of Parc Güell in Barcelona (Figure **G**). But "what if" he had made this Catalonian candy dispenser?

The PEZ dispenser was introduced in the

E

F

1940s, 20 years after Gaudí's death, so this sweet treat toy is clearly a fake. Inspired by the unique chimney designs of Casa Mila, I sculpted some rigid urethane foam to make a parody PEZ head design. Vacuum-formed PETE plastic completes the blister card package. (See makezine.com/projects/kitchen-floor-vacuum-former to build your own Kitchen Floor Vacuum Former, and see *Make:* Volume 74 for details on sculpting and painting foam, in another fun fake toy: Roth's Orbitron as a Hot Wheels car.)

Salvador Dalí, another Catalonian artist, is best known for his surrealist paintings featuring dream-like landscapes and improbable objects: lobster telephones, drooping clocks, and ambiguous optical illusions. Dalí also used his flamboyant talents and personality to make films, fashions, and even advertising for prosaic products like Alka-Seltzer and Braniff International Airways. But this fun fake version of a classic novelty is only a dream (Figure **H**).

Silly Putty a la Dalí is just too funny to resist! The putty picks up a penciled clock face, reversed, to ultimately read right (Figure **I**). Sagging shapes of the stretched putty and a gold painted egg/container are right from Dalí's imagination. I added some novelty plastic ants for the final surrealistic touch (Figure **J**). ✏

1+2+3 Plastic Bag Ice Cream

Written by Maker Camp

TIME REQUIRED: 30 Minutes

DIFFICULTY: Easy

COST: $0–$30

MATERIALS
» **For each serving:**
- Milk, ¼ cup
- Heavy cream, ¼ cup
- Sugar, 1 Tbsp
- Vanilla, ¼ tsp
- Ice, 4 cups
- Salt, 4 Tbsp

» **Quart zip-top freezer bags**
» **Gallon zip-top freezer bags**

Canva

If you shake cream and ice long enough, it turns into ice cream. Sweet! But just how much shaking, and for how long?

❶ MIX YOUR INGREDIENTS

Pour ¼ cup milk, ¼ cup heavy cream, 1 Tbsp sugar, and ¼ tsp vanilla into a quart zip-top freezer bag. Seal the quart bag.

Now place your ice cream bag into a gallon zip-top freezer bag with 4 cups of ice and 4 Tbsp of salt. Seal the gallon bag with everything inside.

Set aside this first bag of ingredients and ice, and prepare two more the same way.

❷ WHOLE LOTTA SHAKIN'

Once prepared, gently shake the second bag for 10 minutes.

Shake the third bag vigorously for 10 minutes. Don't shake the first one at all.

Pour or scoop out the contents of the three bags and compare the results.

MAKER CAMP is our *free* youth program using science, technology, engineering, art, and math (STEAM) to create, build, and discover. Register now at makercamp.com.

❸ FAST AND SLOW

How did your shaking affect the overall completeness (how frozen it is) and smoothness of your ice cream? Which technique created the smoothest ice cream?

WHAT MAKES ICE CREAM CREAMY?

This ice cream is made of sugar, fat, ice crystals, and air. Ice cream's creaminess depends on the size of the ice crystals that form during freezing — the more you shake, the smaller the ice crystals become and the more air is incorporated into the ice cream. Doing both makes for a creamier cream!

WHY DOES SHAKING SOME DAIRY IN A BAG MAKE ICE CREAM?

Ice cream is an *emulsion*, which means small droplets of one liquid dispersed or spread throughout another. Think salad dressing: Oil and vinegar don't dissolve, but they can disperse into an emulsion with the help of a whisk. When you shake the bag, it emulsifies the ice cream, dispersing the ice crystals, fat molecules, and air.

Why salt? Salt lowers the melting temperature of the ice; the colder the icy solution around the ice cream, the faster the cream freezes. ⊘

Go back to a classic...
and beyond the banana piano!

The possibilities are limitless with Makey Makey.®
Check out our new Sampler App plus an
ever-growing catalogue of project ideas and
educational resources at **makeymakey.com**.

MAKE THAT LASER BLAZE

What are the best DIY upgrades to add to your cheap K40 laser cutter? We break it down

Written and photographed by Tim Deagan

TIM DEAGAN loves combining digital fabrication methods with old-school making and crafting techniques in Austin, Texas.

Digital fabrication tools can run from hundreds to many tens of thousands of dollars. For makers interested in a really useful tool at a relatively low initial cost with tremendous upgrade potential, the generic "K40" type 40W CO_2 laser platform is hard to beat.

Significantly more powerful than diode lasers, K40 lasers can cut wood, leather, acrylic, paper, and a range of other materials (generally up to about ¼" thick, though that varies by material). It will engrave on many more types of material, though struggles with metal engraving. While more powerful (and expensive) lasers can outperform these capabilities, the K40 can be significantly improved with end-user upgrades. In almost all cases, applying every upgrade available to a K40 still results in a less expensive tool than moving up to a 50W or 60W laser.

For makers ready to move beyond diode laser solutions, the K40 is a fun, affordable platform ready for modification. Here are my recommendations, based on upgrades I've done to mine.

① OMTECH K40

If you're going to upgrade a K40, you need to start with a K40. I got mine from OMTech, a U.S.-based company that adds value to the purchase of the general K40 model (for a price of course) by checking the quality of build, components, etc. Cheaper, nearly identical units are directly importable, but often have component or build challenges. This unit is absolutely functional out of the box, but the platform is also a great base for upgrades.

- **Price:** $450, **Return on investment (ROI):** 8
- **Ease of installation (1=poor, 10=excellent):** 8
- **Link to part:** ebay.com/itm/264443204036
- **Instructions:** k40.se/k40-laser-setup
- **Recommendation: TOTALLY NECESSARY!**

② 3MA ANALOG METER

This is an essential upgrade. The stock units report power as a percentage of power supply capacity, but this varies from machine to machine. To be able to use cutting or engraving settings from other people, the actual amperage needs to be used, and a meter is needed to be able to discern it.

- **Price:** $10, **ROI:** 10
- **Ease of setup:** 4
- **Link to part:** smile.amazon.com/gp/product/B07DH77XBS
- **Instructions:** k40.se/k40-laser-electronics/install-ma-meter
- **Recommendation: MUST DO**

③ SAFETY GLASSES

CO_2 lasers are not visible to the human eye. You can't tell if backscatter from the beam is causing damage until after it's occurred. The lid of the K40 has a tinted panel that serves as a filter, but whenever you're aligning mirrors or doing other operations that need to fire the laser while the lid is up, appropriate eye protection is essential. The wavelength that the K40 CO_2 laser emits ranges from 10,400–10,600nm; glasses rated at 10,600nm are the most readily available.

- **Price:** $45, **ROI:** 10
- **Ease of installation:** 10
- **Recommendation: MUST DO**

④ LIGHTBURN SOFTWARE

LightBurn is a full-featured laser controlling package. While a variety of functional freeware packages like K40Whisperer are totally usable, LightBurn offers many more features that make it worth the license fee. LightBurn requires an upgraded controller board (e.g. the Cohesion3D LaserBoard listed below) to work with a K40.

- **Price:** $60, **ROI:** 8
- **Ease of installation:** 10
- **Link to part:** cohesion3d.com/shop/software/lightburn-software
- **Instructions:** lasercutting.avataar120.com/en/2021/08/01/installing-lightburn-and-setting-it-up-in-3-steps
- **Recommendation: NICE TO DO**

⑤ COHESION3D LASERBOARD

The stock K40 controller board cannot vary laser power, so it can't do actual grayscale images — it only does dither-type "grays." It also limits the software that can be used. A variety of after-market controllers are available, with the Cohesion3D LaserBoard arguably the

most popular. This Smoothie firmware-based board supports external displays/controllers, grayscaling, rotary axis tools, air-assist relays, and advanced software packages like LightBurn.

- **Price:** $230, **ROI:** 7
- **Ease of installation:** 4
- **Link to part:** cohesion3d.com/shop/controllers/cohesion3d-laserboard
- **Instructions:** cohesion3d.com/knowledgebase/k40-with-m2nano-cohesion3d-laserboard-installation
- **Recommendation: NICE TO DO**

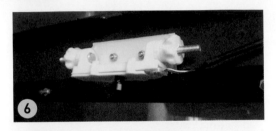

⑥ LIGHTBURN CAMERA

This camera mounts in the lid and works with the LightBurn software to allow a picture of the material on the cutting bed to be used in the software to orient and position the images and lines to be engraved or cut. It's not essential, but once you use it, you'll never ever want to go back to other registration methods. It comes as a thin PCB; I printed an enclosure for mine.

- **Price:** $80, **ROI:** 8
- **Ease of installation:** 5
- **Link to part:** cohesion3d.com/shop/peripherals/lightburn-camera/official-lightburn-camera
- **Instructions:** lightburnsoftware.github.io/NewDocs/UsingACamera.html
- **Recommendation: NICE TO DO**

⑦ LID STAND

The K40 lid will rest in the "far back" position (about 15° past vertical,) but given a bump or strong wind, it can come crashing down. This isn't a big problem, but using a lid stand definitely feels safer.

- **Price:** $0, **ROI:** 10
- **Ease of installation:** 10
- **Link to part:** Scrap material and 3D printed (or laser cut!) parts
- **Recommendation: NICE TO DO (MUST DO if using LightBurn camera)**

NOTE: To use the LightBurn camera, you need to have the lid positioned consistently at about 60% of full open. I addressed this by cutting a piece of ½" angle aluminum and 3D printing ends that fit on the K40 base and lid. When I wedge open the lid with it, I know it's in the correct position for the camera.

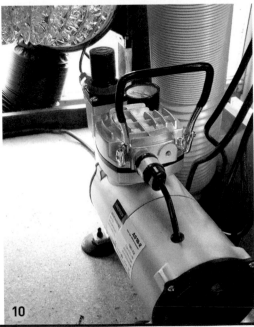

10

⑧ CROSSHAIR LASER SIGHTS

Without something like the LightBurn camera, laser crosshairs to visually display where the beam will hit are a great aid in positioning material.

- **Price:** $8, **ROI:** 7
- **Ease of installation:** 4
- **Link to part:** smile.amazon.com/gp/product/B07L428BDG
- **Recommendation: NICE TO DO**

⑨ AIR ASSIST HEAD

This device replaces the stock head (or just the bottom of it), allowing a source of compressed air to shoot down along the laser beam. This suppresses flames and the resulting char that the laser causes with many materials.

- **Price:** $25, **ROI:** 8
- **Ease of installation:** 9
- **Link to part:** smile.amazon.com/gp/product/B0094WLPYK
- **Instructions:** youtube.com/watch?v=8S2aNjPM7P8
- **Recommendation: SHOULD DO**

⑩ AIRBRUSH COMPRESSOR

For air, there are a variety of heavy duty aquarium pumps and other options. I bought an airbrush compressor since it is also useful for painting.

- **Price:** $65, **ROI:** 5
- **Ease of installation:** 7
- **Link to part:** smile.amazon.com/gp/product/B07VSFZVRH
- **Recommendation: MUST DO (some solution if using air-assist)**

⑪ EXHAUST SYSTEM

The stock exhaust fan is just barely able to clear smoke and fumes from the laser. A more powerful fan is a big win to keep from stinking up your shop and accumulating residue on your mirrors and lens.

- **Price:** $60, **ROI:** 8
- **Ease of installation:** 7
- **Links to parts:**
  smile.amazon.com/gp/product/B003NE59HE
  smile.amazon.com/gp/product/B01M7S46YZ
  smile.amazon.com/gp/product/B07WRDTX4C
  smile.amazon.com/gp/product/B01M7S46YZ
- **Recommendation: SHOULD DO**

11

12

⑫ Z-HEIGHT LAB JACK

By removing the static deck provided with the K40, you can put a height-adjustable table in its place. This is important because the focal point of the laser never changes; instead, you move the material up and down to accommodate different thicknesses. This jack's table surface is only 4"×4", so it needs something on top of it. It's easy to accidentally tilt the top, so use a level to make sure it's true to the laser's X and Y axes.

- **Price:** $20, **ROI:** 8
- **Ease of installation:** 6
- **Link to part:** smile.amazon.com/gp/product/B07KDXJGX9
- **Recommendation: SHOULD DO**

⑬ LENS

This is a drop-in replacement for the stock focus lens that has better coatings and performance.

- **Price:** $25, **ROI:** 7
- **Ease of installation:** 9
- **Link to part:** smile.amazon.com/gp/product/B01DP2HMPK
- **Instructions:** youtube.com/watch?v=5AptkbV-v6E
- **Recommendation: SHOULD DO**

⑭ MIRRORS

Drop-in replacements for the stock mirrors that have better coatings and performance.

- **Price:** $33, **ROI:** 7
- **Ease of installation:** 9
- **Link to part:** smile.amazon.com/gp/product/B01DP2HV5Q
- **Instructions:** youtube.com/watch?v=5AptkbV-v6E
- **Recommendation: SHOULD DO**

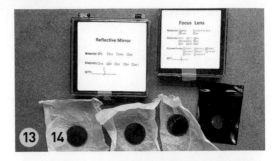

⑮ CW 5200 WATER CHILLER

The CO_2 laser tube is water cooled. Operating it above its desired temperature (around 77°F/25°C) rapidly degrades its performance and longevity. Routing the water through buckets of ice water is a cheap option, but difficult to maintain in hot climates for long runs. I opted for a significantly oversized active water cooler since my shop gets close to 100°F/38°C in the summer. Be careful of cheaper thermolysis-type "coolers" that only pass the water through a radiator with a fan, limiting the cooling ability depending on ambient temperatures.

- **Price:** $450, **ROI:** 6
- **Ease of installation:** 8
- **Link to part:** ebay.com/itm/254957802153?var=554846117300
- **Recommendation: MUST DO (something for cooling); NICE TO DO (active chiller)**

⑯ DIY ROTARY AXIS

The K40 has a very limited Z-axis space. Most commercial rotary axis tools (that allow you to engrave or cut on cylindrical objects) will not fit inside and still keep your workpiece at or below the necessary focal height. A 3D printed, low-profile rotary axis is a fun project and can be the best use of space in the tool.

- **Price:** $35, **ROI:** 6
- **Ease of installation:** 4
- **Link to part:** thingiverse.com/thing:3174149
- **Recommendation: NICE TO DO**

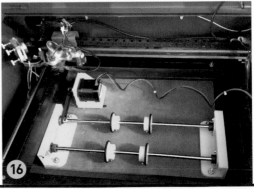

⑰ HONEYCOMB BED

When the laser finally punches through the material during cutting, it hits whatever the material was sitting on. If it's a material that burns, then smoke or stains may accrue on the bottom of the workpiece. Honeycomb is one way to address parts of the problem by providing a deck that is largely open under the workpiece. If the base under the honeycomb is closed, the smoke that collects in the cells can ignite and add oddly patterned char, so it's not perfect for all materials.

- **Price:** $18, **ROI:** 5
- **Ease of installation:** 7
- **Link to part:** etsy.com/listing/924015043/ aluminum-honeycomb-plate-300x200x10mm
- **Recommendation: NICE TO DO**

⑱ Z-AXIS TABLE

An adjustable-height table that evenly raises and lowers the deck while keeping it level is a big enhancement. Thingiverse has a number of models for both stepper-motor and manually driven tables.

- **Price:** Varies, **ROI:** 8
- **Ease of installation:** 4
- **Link to part:** Search "K40 table" (there are dozens) on Thingiverse or your favorite 3D repository. Add assorted nuts, bolts, shafts, etc. as directed.
- **Recommendation: NICE TO DO**

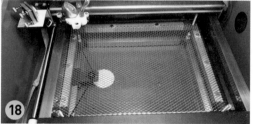

⑲ GRAPHIC LCD

This is an external display and knob/button that provides the same kind of on-machine information and control that you get on newer 3D printers.

- **Price:** $45, **ROI:** 4
- **Ease of installation:** 9
- **Link to part:** cohesion3d.com/shop/peripherals/ graphic-lcd-control-panel-with-adapter
- **Instructions:** cohesion3d.com/knowledgebase/ graphic-lcd-overview
- **Recommendation: NICE TO DO**

SUMMARY

I paid $454 for the OMTech K40, and $1,203 for the upgrades. Including sales tax in Austin, Texas, my final cost for this whole system was $1,794.

I was skeptical that a 40W laser cutter/

engraver would be a practical tool. But despite its limited workspace (12"×8", or about 300mm× 200mm) the range of materials it works with combined with the ease of setup and use has made my K40 a go-to tool, especially for creative projects, in my shop. Plus, it was tons of fun to mod and upgrade! ◐

Sondehub view of weather balloons above Europe.

PLANE SPOTTING (AND MORE!)

Written by onemindisbuddha

Put your spare gear to use as a feeder for **community-based data exchanges**

If you've ever used a flight-tracking app or looked up local weather on Weather Underground, you've likely accessed data supplied, in part, by volunteer contributors. Crowd-sourced *data exchanges* collect this information and share the pooled results online via interactive maps. The data ranges from radio-signal telemetry for airplanes, ships, weather balloons, or satellites, to seismic measurements, to air and weather quality. The end result is providing a useful service a larger community through apps or websites that may require a paid membership, or may employ the support of advertising. In return for the time and money spent to generate and feed data into their commercial exchange, these outlets sometimes offer the participants enterprise-level access to their apps and services, which can otherwise be quite costly. Some feeders might also simply want to have their data viewable through a specific provider's UX, or enjoy contributing scientific data to a project alongside an international community.

The devices that feed these are sometimes proprietary, provided or sold by commercial data exchanges, but are more commonly something the participants make. For radio data, a typical configuration would be connecting a Raspberry Pi with a customized OS to an inexpensive RTL-SDR dongle and a handmade antenna. Even the early RasPi models work well, although it can be advantageous to have Wi-Fi — built-in or via dongle — for optimal tracker placement

With help from others online, I've assembled the following listing of many data exchanges in which you can participate. There are likely others, and note that this list doesn't contain exchanges requiring an amateur radio license such as APRS, PSKReporter, or WSPRnet. I encourage you to pull out a spare Pi and grab a USB-equipped SDR (they cost around $25) to play around with any that might catch your eye. Have fun!

Flightradar24 users track data on all types of flights.

SafeCast sensor map of radiation levels in Fukushima.

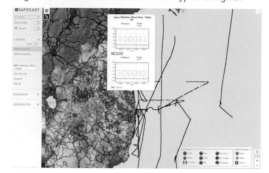
Raspberry Pi connected to a SDR dongle and DIY antenna.

Weather station feeding data to Weather Underground.

ONEMINDISBUDDHA is a 38-year Japan resident, *Make:* fan, and a radio and 3D printing enthusiast. You can find him on Reddit under the same name.

DIY DATA EXCHANGES

TYPE	NAME/LINK	USING	DEVICES
AIR TRAFFIC	ADS-B Exchange adsbexchange.com	ADS-B	Raspberry Pi + RTL-SDR
AIR TRAFFIC	Airframes app.airframes.io	ACARS	Raspberry Pi + RTL-SDR
AIR TRAFFIC	FlightAware flightaware.com	ADS-B	Raspberry Pi + RTL-SDR
AIR TRAFFIC	Flightradar24 flightradar24.com	ADS-B	Raspberry Pi + RTL-SDR
AIR TRAFFIC	OGN (Open Glider Network) glidernet.org	FLARM, OGN	Raspberry Pi + RTL-SDR
AIR TRAFFIC	OpenSky opensky-network.org	ADS-B	Raspberry Pi + RTL-SDR
AIR TRAFFIC	Plane Finder planefinder.net	ADS-B	Raspberry Pi + RTL-SDR *
AIR TRAFFIC	RadarVirtuel radarvirtuel.com	ADS-B, FLARM, ATC, ACARS	Raspberry Pi + RTL-SDR
AIR TRAFFIC	Thebaldgeek thebaldgeek.github.io	ACARS, etc.	Raspberry Pi + RTL-SDR
LIGHTNING	Blitzortung (blitzortung.org/en/live_lightning_maps.php); feeds LightningMaps lightningmaps.org	N/A	Proprietary lightning detection devices
MARINE TRAFFIC	AISHub aishub.net	AIS	Raspberry Pi + RTL-SDR
MARINE TRAFFIC	Pocket Mariner pocketmariner.com	AIS	Supplied proprietary, Raspberry Pi + RTL-SDR **
MARINE TRAFFIC	Marine Traffic marinetraffic.com	AIS	Supplied proprietary, Raspberry Pi + RTL-SDR
MARINE TRAFFIC	Ship Finder shipfinder.co	AIS	Raspberry Pi + RTL-SDR
MARINE TRAFFIC	VesselFinder vesselfinder.com	AIS	Raspberry Pi + RTL-SDR **
RADIATION, AIR QUALITY	SafeCast safecast.org	Sensors	Proprietary
RADIO	Broadcastify broadcastify.com/listen	Radio feeds (police/EMS/etc.)	PC + radio
RADIO	OpenMHz openmhz.com	Radio feeds (police, fire) (U.S.)	PC + RTL-SDR
SATELLITES	R2server r2server.ru	APT, LRPT, FSK, etc.	Raspberry Pi + RTL-SDR
SATELLITES	SatNOGS network.satnogs.org	VHF/UHF, mixed	Raspberry Pi + RTL-SDR
SATELLITES	TinyGS tinygs.com	LoRa	LoRa capable ESP32
SDR	RX-TX rx-tx.info/map-sdr-points	AM/FM/CW/LSB/USB/etc.	RTL-SDR w/ WebSDR, OpenWebRX, or KiwiSDR
SEISMIC ACTIVITY	RaspberryShake raspberryshake.org	ADS-B	Proprietary
TRAIN TRAFFIC (U.S.)	ATCSMon (groups.io/g/ATCSMonitor); feeds TrainMon5 trainmon5.com/About.aspx	ATCS, ARES, etc.	Raspberry Pi + RTL-SDR
WEATHER	Weather Underground wunderground.com/pws/overview	N/A	Commercial weather stations + bridge
WEATHER BALLOONS	SQ6KXY Radiosonde Tracker radiosondy.info	Various (RS41, etc.)	Raspberry Pi + RTL-SDR
WEATHER BALLOONS	SondeHub sondehub.org	Various (RS41, etc.)	Raspberry Pi + RTL-SDR; LoRa capable ESP32

* These also supply proprietary receivers to plug gaps in coverage.

** Also accepts NMEA-0183 messages via UDP socket that is relatively simple to set up from generic software.

Maritime activity, tracked by users of Ship Finder.

GLOSSARY

APRS: Automatic Packet Reporting System — an amateur radio system for real time communication of coordinates, weather station telemetry, text messages, etc.

WSPR: Weak Signal Propagation Reporter — a protocol for testing propagation paths on the medium (300-3,000kHz) and high (3-30MHz) frequency radio bands.

ADS-B: Automatic Dependent Surveillance-Broadcast — a flight data transmission system with wide global adoption.

ACARS: Aircraft Communication Addressing and Reporting System — a system for transmission of short messages between aircraft and ground stations via radio or satellite.

FLARM: portmanteau of "flight" and "alarm" — a proprietary system which alerts pilots to potential collisions between aircraft, most commonly used in light aircraft such as gliders.

OGN: Open Glider Network — a unified tracking system for gliders, drones, and other aircraft.

ATC: Air Traffic Control — "airband" radio communication between air traffic control and aircraft.

AIS: Automatic Identification System — coastal tracking system for ships.

APT: Automatic Picture Transmission — low-resolution analog image transmission system used by weather satellites.

LRPT: Low-Rate Picture Transmission — digital image transmission system used by weather satellites.

VHF/UHF: Very High Frequency / Ultra High Frequency — radio bands in which data is broadcast. VHF: 30–300MHz, UHF: 300MHz–3GHz.

LoRa: portmanteau of "long" and "range" — proprietary low-power wide-area network modulation commonly used in IoT, with recent use in satellites.

AM/FM/CW/LSB/USB/FSK/PSK: Amplitude Modulation / Frequency Modulation / Continuous Wave / Lower Side Band / Upper Side Band / Frequency-Shift Keying / Phase Shift Keying — modes of radio communication.

ATCS: Automated Train Control System — a train data transmission system in use in the U.S.

ARES: Advanced Railroad Electronics System — a version of ATCS specific to Burlington Northern Railroad. Not to be confused with the Amateur Radio Emergency Service.

RS41: Data format for the RS41 weather balloon manufactured by Vaisala. One of 10 weather balloon data formats tracked. ◐

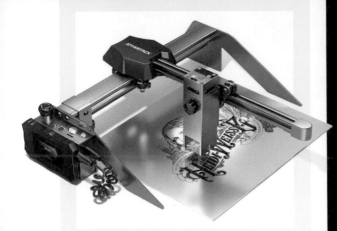

Atomstack P9 Laser Engraver

$499 atomstack.net

The Atomstack P9 is the latest in a series of desktop diode-based laser engravers that have been hitting the market. This has a few standout features, like tilt detection, which will disable the laser if the machine gets knocked over, and a nice removable touch screen for control. This is a very bare-bones machine though. You're going to manually adjust your focus and practice your patience on endeavors like engraving an image. Atomstack also falls prey to the tendency to market overall machine power — they sell this as a 50-watt laser on their website, even though the actual laser output is only 10 watts! At $499, it's in the same price range as a cheap K40 which makes this kind of a hard sell. If portability is important, this one definitely comes out on top.

Mingda Magician X

$329 3dmingdaofficial.com

The Magician X 3D printer is packed with features, such as auto bed leveling and a filament runout sensor. The dual Z-axis screws seem like they provide a very solid platform for some nice prints, and our test found the quality out of the box to be acceptable. It was also very easy to assemble straight out of the box; getting it together took just 5–10 minutes.

A couple things to note: The machine we got didn't ship with a pre-configured slicer. Pros won't have a problem creating their own config, but beginners will be lost and confused here. The auto bed leveling that they tout seems quite capable, however there's no fallback way to manually level; if your auto level system fails, you're just out of luck. I learned this the hard way — my first unit did fail! Their customer support replaced it very quickly, though.

This machine does have a glass bed. If you're a fan of them, you'll be happy here.

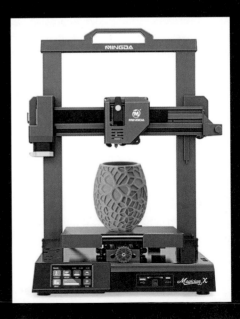

Craft Circuits on Paper with Circuit Stickers!

Make STEAM Learning Shine!

Dremel Multi–Max Oscillating Tool

$159 dremel.com/us/en

An oscillating tool is one of those devices that you might rarely need, but the second you find a use for one, you're immensely happy you have one on hand. Dremel's latest offering, the Multi-Max 20V, brings a few nice features to the table. The 20V battery supplies enough power to keep this thing from feeling anemic, but is also quite low profile, so it won't get in the way when you're reaching into awkward spaces to cut. Dremel has also angled the cutting head a bit, which seems like an obvious upgrade but this is the first time I've seen it. Swapping tool heads is super easy with their tool-less accessory change design. It's a good one to keep in your toolbox for those moments you might not anticipate.

Voxelab Aquila S2

$279 voxelab3dp.com

Voxelab is the budget arm of Flashforge, a popular 3D printer brand, and the price on this machine really drives home the meaning of "budget."

Assembling this printer was slightly more work than I'm used to with cheap systems, but it wasn't as complex as a full kit either. Unlike many modern printers, there's no auto bed leveling or filament runout sensor, but my initial experiences with it have been surprisingly good. After manually leveling the (removable and flexible) print bed, it has been printing quite nicely without issue. The Aquila S2 can get up to 300°C, which is actually a big selling point, especially for such a cheap printer.

Tell us about your faves: editor@makezine.com

DETERMI**NATION**

Determination seems like the best word to describe the collective response of Ukrainians as they fight for their country, for their home, and their freedom. Volodymyr Zelenskyy, President of Ukraine, described it as "people power."

Volunteers in workshops, like the one shown above, are contributing to the defense of Ukraine. They are welding barriers for tanks and other vehicles. The incredible photograph was taken by Dmytro Kovalenko, a video producer and photographer who is helping international journalists cover the war in Ukraine.

The photo appeared on the home page of Maker Hub (makerhub.org) to help raise support for makers and makerspaces in the Ukraine. I interviewed Yuri Vlasyuk and Svitlana Bovkun, who live in Kyiv for an episode of *Make:cast* called "Determination" (makezine.com/2022/03/25/determination-makers-in-ukraine). They have co-produced 15 Maker Faires in five cities in Ukraine, dating back to 2015. "We understand that there is no one who will fight for our country," Yuri told me. "We have to fight by ourselves."

If you'd like to help, go to makerhub.org/support.

—*Dale Dougherty*